CAMBRIDGE LIBRARY COLLECTION

Books of enduring scholarly value

Life Sciences

Until the nineteenth century, the various subjects now known as the life sciences were regarded either as arcane studies which had little impact on ordinary daily life, or as a genteel hobby for the leisured classes. The increasing academic rigour and systematisation brought to the study of botany, zoology and other disciplines, and their adoption in university curricula, are reflected in the books reissued in this series.

Wood and Garden

Gertrude Jekyll (1843–1932) was one of the most influential garden designers of the late nineteenth and early twentieth centuries. Skilled as a painter and in many forms of handicrafts, she found her metier in the combination of her artistic skills with considerable botanical knowledge. Having been collecting and breeding plants, including Mediterranean natives, since the 1860s, she began writing for William Robinson's magazine, *The Garden*, in 1881, and together they are regarded as transforming English horticultural method and design: Jekyll herself received over 400 design commissions in Britain, and her few surviving gardens are treasured today. Like Robinson's, her designs were informal and more natural in style than earlier Victorian fashions. In this, the first of fourteen books, published in 1899, she stresses the importance of being inspired by nature, and sums up her philosophy of gardening: 'planting ground is painting a landscape with living things'.

Cambridge University Press has long been a pioneer in the reissuing of out-of-print titles from its own backlist, producing digital reprints of books that are still sought after by scholars and students but could not be reprinted economically using traditional technology. The Cambridge Library Collection extends this activity to a wider range of books which are still of importance to researchers and professionals, either for the source material they contain, or as landmarks in the history of their academic discipline.

Drawing from the world-renowned collections in the Cambridge University Library, and guided by the advice of experts in each subject area, Cambridge University Press is using state-of-the-art scanning machines in its own Printing House to capture the content of each book selected for inclusion. The files are processed to give a consistently clear, crisp image, and the books finished to the high quality standard for which the Press is recognised around the world. The latest print-on-demand technology ensures that the books will remain available indefinitely, and that orders for single or multiple copies can quickly be supplied.

The Cambridge Library Collection will bring back to life books of enduring scholarly value (including out-of-copyright works originally issued by other publishers) across a wide range of disciplines in the humanities and social sciences and in science and technology.

Wood and Garden

*Notes and Thoughts, Practical and Critical,
of a Working Amateur*

GERTRUDE JEKYLL

CAMBRIDGE
UNIVERSITY PRESS

CAMBRIDGE UNIVERSITY PRESS

Cambridge, New York, Melbourne, Madrid, Cape Town,
Singapore, São Paolo, Delhi, Tokyo, Mexico City

Published in the United States of America by Cambridge University Press, New York

www.cambridge.org
Information on this title: www.cambridge.org/9781108037198

© in this compilation Cambridge University Press 2011

This edition first published 1899
This digitally printed version 2011

ISBN 978-1-108-03719-8 Paperback

WOOD AND GARDEN

Frontispiece.

WOOD AND GARDEN

NOTES AND THOUGHTS, PRACTICAL AND CRITICAL, OF A WORKING AMATEUR

BY

GERTRUDE JEKYLL

*With Illustrations from Photographs
by the Author*

LONGMANS, GREEN, AND CO.
39 PATERNOSTER ROW, LONDON
NEW YORK AND BOMBAY
1899

Printed by BALLANTYNE, HANSON & Co.
At the Ballantyne Press

PREFACE

FROM its simple nature, this book seems scarcely to need any prefatory remarks, with the exception only of certain acknowledgments.

A portion of the contents (about one-third) appeared during the years 1896 and 1897 in the pages of the *Guardian,* as "Notes from Garden and Woodland." I am indebted to the courtesy of the editor and proprietors of that journal for permission to republish these notes.

The greater part of the photographs from which the illustrations have been prepared were done on my own ground—a space of some fifteen acres. Some of them, owing to my want of technical ability as a photographer, were very weak, and have only been rendered available by the skill of the reproducer, for whose careful work my thanks are due.

A small number of the photographs were done for reproduction in wood-engraving for Mr. Robinson's *Garden, Gardening Illustrated,* and *English Flower Garden.* I have his kind permission to use the original plates.

G. J.

CONTENTS

LIST OF ILLUSTRATIONS

WOOD AND GARDEN

CHAPTER I

INTRODUCTORY

THERE are already many and excellent books about gardening; but the love of a garden, already so deeply implanted in the English heart, is so rapidly growing, that no excuse is needed for putting forth another.

I lay no claim either to literary ability, or to botanical knowledge, or even to knowing the best practical methods of cultivation; but I have lived among outdoor flowers for many years, and have not spared myself in the way of actual labour, and have come to be on closely intimate and friendly terms with a great many growing things, and have acquired certain instincts which, though not clearly defined, are of the nature of useful knowledge.

But the lesson I have thoroughly learnt, and wish to pass on to others, is to know the enduring happiness that the love of a garden gives. I rejoice when I see any one, and especially children, inquiring about flowers, and wanting gardens of their own, and carefully working

A

in them. For the love of gardening is a seed that once sown never dies, but always grows and grows to an enduring and ever-increasing source of happiness.

If in the following chapters I have laid special stress upon gardening for beautiful effect, it is because it is the way of gardening that I love best, and understand most of, and that seems to me capable of giving the greatest amount of pleasure. I am strongly for treating garden and wooded ground in a pictorial way, mainly with large effects, and in the second place with lesser beautiful incidents, and for so arranging plants and trees and grassy spaces that they look happy and at home, and make no parade of conscious effort. I try for beauty and harmony everywhere, and especially for harmony of colour. A garden so treated gives the delightful feeling of repose, and refreshment, and purest enjoyment of beauty, that seems to my understanding to be the best fulfilment of its purpose; while to its diligent worker its happiness is like the offering of a constant hymn of praise. For I hold that the best purpose of a garden is to give delight and to give re- freshment of mind, to soothe, to refine, and to lift up the heart in a spirit of praise and thankfulness. It is certain that those who practise gardening in the best ways find it to be so.

But the scope of practical gardening covers a range of horticultural practice wide enough to give play to every variety of human taste. Some find their greatest pleasure in collecting as large a number as possible of

all sorts of plants from all sources, others in collecting
them themselves in their foreign homes, others in making
rock-gardens, or ferneries, or peat-gardens, or bog-gardens,
or gardens for conifers or for flowering shrubs, or special
gardens of plants and trees with variegated or coloured
leaves, or in the cultivation of some particular race or
family of plants. Others may best like wide lawns with
large trees, or wild gardening, or a quite formal garden,
with trim hedge and walk, and terrace, and brilliant
parterre, or a combination of several ways of gardening.
And all are right and reasonable and enjoyable to
their owners, and in some way or degree helpful to
others.

The way that seems to me most desirable is again
different, and I have made an attempt to describe it
in some of its aspects. But I have learned much, and
am always learning, from other people's gardens, and
the lesson I have learned most thoroughly is, never
to say "I know"—there is so infinitely much to
learn, and the conditions of different gardens vary so
greatly, even when soil and situation appear to be
alike and they are in the same district. Nature is
such a subtle chemist that one never knows what
she is about, or what surprises she may have in store
for us.

Often one sees in the gardening papers discussions
about the treatment of some particular plant. One
man writes to say it can only be done one way,
another to say it can only be done quite some other

way, and the discussion waxes hot and almost angry, and the puzzled reader, perhaps as yet young in gardening, cannot tell what to make of it. And yet the two writers are both able gardeners, and both absolutely trustworthy, only they should have said, "In my experience *in this place* such a plant can only be done in such a way." Even plants of the same family will not do equally well in the same garden. Every practical gardener knows this in the case of strawberries and potatoes; he has to find out which kinds will do in his garden; the experience of his friend in the next county is probably of no use whatever.

I have learnt much from the little cottage gardens that help to make our English waysides the prettiest in the temperate world. One can hardly go into the smallest cottage garden without learning or observing something new. It may be some two plants growing beautifully together by some happy chance, or a pretty mixed tangle of creepers, or something that one always thought must have a south wall doing better on an east one. But eye and brain must be alert to receive the impression and studious to store it, to add to the hoard of experience. And it is important to train oneself to have a good flower-eye; to be able to see at a glance what flowers are good and which are unworthy, and why, and to keep an open mind about it; not to be swayed by the petty tyrannies of the "florist" or show judge; for, though some part of his judgment may be sound, he is himself a slave to rules, and must go by points which are defined arbitrarily and rigidly, and have reference mainly to the show-table, leaving out of account, as if unworthy of consideration, such matters as gardens and garden beauty, and human delight, and sunshine, and varying lights of morning and evening and noonday. But many, both nursery-men and private people, devote themselves to growing and improving the best classes of hardy flowers, and we can hardly offer them too much grateful praise, or do them too much honour. For what would our gar-dens be without the Roses, Pæonies, and Gladiolus of France, and the Tulips and Hyacinths of Holland, to say nothing of the hosts of good things raised by our home growers, and of the enterprise of the great firms whose agents are always searching the world for garden treasures ?

Let no one be discouraged by the thought of how much there is to learn. Looking back upon nearly thirty years of gardening (the earlier part of it in groping ignorance with scant means of help), I can remember no part of it that was not full of pleasure and encouragement. For the first steps are steps into a delightful Unknown, the first successes are victories all the happier for being scarcely expected, and with the growing knowledge comes the widening outlook, and the comforting sense of an ever-increasing gain of critical appreciation. Each new step becomes a little surer, and each new grasp a little firmer, till, little by little, comes the power of intelligent combination, the nearest thing we can know to the mighty force of creation.

And a garden is a grand teacher. It teaches patience and careful watchfulness; it teaches industry and thrift; above all, it teaches entire trust. " Paul planteth and Apollos watereth, but God giveth the increase." The good gardener knows with absolute certainty that if he does his part, if he gives the labour, the love, and every aid that his knowledge of his craft, experience of the conditions of his place, and exercise of his personal wit can work together to suggest, that so surely as he does this diligently and faithfully, so surely will God give the increase. Then with the honestly-earned success comes the consciousness of encouragement to renewed effort, and, as it were, an echo of the gracious words, " Well done, good and faithful servant."

CHAPTER II

JANUARY

Beauty of woodland in winter—The nut-walk—Thinning the over-
growth—A nut nursery—*Iris stylosa*—Its culture—Its home in
Algeria Discovery of the white variety—Flowers and branches
for indoor decoration.

A HARD frost is upon us. The thermometer registered
eighteen degrees last night, and though there was only
one frosty night next before it, the ground is hard
frozen. Till now a press of other work has stood in the
way of preparing protecting stuff for tender shrubs, but
now I go up into the copse with a man and chopping
tools to cut out some of the Scotch fir that are
beginning to crowd each other.

How endlessly beautiful is woodland in winter!
To-day there is a thin mist; just enough to make
a background of tender blue mystery three hundred
yards away, and to show any defect in the grouping
of near trees. No day could be better for deciding
which trees are to come down; there is not too much
at a time within sight; just one good picture-full and
no more. On a clear day the eye and mind are dis-
tracted by seeing away into too many planes, and it is

much more difficult to decide what is desirable in the
way of broad treatment of nearer objects.

The ground has a warm carpet of pale rusty fern;
tree-stem and branch and twig show tender colour-
harmonies of grey bark and silver-grey lichen, only
varied by the warm feathery masses of birch spray.
Now the splendid richness of the common holly is
more than ever impressive, with its solid masses of
full, deep colour, and its wholesome look of perfect
health and vigour. Sombrely cheerful, if one may
use such a mixture of terms; sombre by reason of the
extreme depth of tone, and yet cheerful from the look
of glad life, and from the assurance of warm shelter
and protecting comfort to bird and beast and neighbouring vegetation. The picture is made complete
by the slender shafts of the silver-barked birches, with
their half-weeping heads of delicate, warm-coloured
spray. Has any tree so graceful a way of throwing
up its stems as the birch? They seem to leap and
spring into the air, often leaning and curving upward
from the very root, sometimes in forms that would
be almost grotesque were it not for the never-failing
rightness of free-swinging poise and perfect balance.
The tints of the stem give a precious lesson in colour.
The white of the bark is here silvery-white and there
milk-white, and sometimes shows the faintest tinge of
rosy flush. Where the bark has not yet peeled, the
stem is clouded and banded with delicate grey, and
with the silver-green of lichen. For about two feet

upward from the ground, in the case of young trees
of about seven to nine inches diameter, the bark is
dark in colour, and lies in thick and extremely rugged
upright ridges, contrasting strongly with the smooth
white skin above. Where the two join, the smooth
bark is parted in upright slashes, through which the
dark, rough bark seems to swell up, reminding one
forcibly of some of the old fifteenth-century German
costumes, where a dark velvet is arranged to rise in
crumpled folds through slashings in white satin. In
the stems of older birches the rough bark rises much
higher up the trunk and becomes clothed with delicate
grey-green lichen.

The nut-walk was planted twelve years ago. There
are two rows each side, one row four feet behind the
other, and the nuts are ten feet apart in the rows.
They are planted zigzag, those in the back rows showing between the front ones. As the two inner rows
are thirteen feet apart measuring across the path, it
leaves a shady border on each side, with deeper bays
between the nearer trees. Lent Hellebores fill one
border from end to end; the other is planted with
the Corsican and the native kinds, so that throughout
February and March there is a complete bit of garden
of one kind of plant in full beauty of flower and
foliage.

The nut-trees have grown into such thick clumps
that now there must be a vigorous thinning. Each
stool has from eight to twelve main stems, the largest

of them nearly two inches thick. Some shoot almost upright, but two or three in each stool spread outward, with quite a different habit of growth, branching about in an angular fashion. These are the oldest and thickest. There are also a number of straight suckers one and two years old. Now when I look at some fine old nut alley, with the tops arching and meeting overhead, as I hope mine will do in a few years, I see that the trees have only a few stems, usually from three to five at the most, and I judge that now is the time to thin mine to about the right number, so that the strength and growing power may be thrown into these, and not allowed to dilute and waste itself in growing extra faggoting. The first to be cut away are the old crooked stems. They grow nearly horizontally and are all elbows, and often so tightly locked into the straighter rods that they have to be chopped to pieces before they can be pulled out. When these are gone it is easier to get at the other stems, though they are often so close together at the base that it is difficult to chop or saw them out without hurting the bark of the ones to be left. All the young suckers are cut away. They are of straight, clean growth, and we prize them as the best possible sticks for Chrysanthemums and potted Lilies.

After this bold thinning, instead of dense thickety bushes we have a few strong, well-branched rods to each stool. At first the nut-walk looks wofully naked, and for the time its pictorial value is certainly lessened;

but it has to be done, and when summer side-twigs have grown and leafed, it will be fairly well clothed, and meanwhile the Hellebores will be the better for the thinner shade.

The nut-catkins are already an inch long, but are tightly closed, and there is no sign as yet of the bright crimson little sea-anemones that will appear next month and will duly grow into nut-bearing twigs. Round the edges of the base of the stools are here and there little branching suckers. These are the ones to look out for, to pull off and grow into young trees. A firm grasp and a sharp tug brings them up with a fine supply of good fibrous root. After two years in the nursery they are just right to plant out.

The trees in the nut-walk were grown in this way fourteen years ago, from small suckers pulled off plants that came originally from the interesting cob-nut nursery at Calcot, near Reading.

I shall never forget a visit to that nursery some six-and-twenty years ago. It was walled all round, and a deep-sounding bell had to be rung many times before any one came to open the gate; but at last it was opened by a fine, strongly-built, sunburnt woman of the type of the good working farmer's wife, that I remember as a child. She was the forewoman, who worked the nursery with surprisingly few hands—only three men, if I remember rightly—but she looked as if she could do the work of " all two men " herself. One of the specialties of the place was a fine breed of mastiffs;

another was an old Black Hamburg vine, that rambled
and clambered in and out of some very old green-
houses, and was wonderfully productive.   There were
alleys of nuts in all directions, and large spreading
patches of palest yellow Daffodils—the double *Narcissus cernuus,* now so scarce and difficult to grow.   Had
I then known how precious a thing was there in fair
abundance, I should not have been contented with the
modest dozen that I asked for.   It was a most pleasant garden to wander in, especially with the old Mr.
Webb who presently appeared.   He was dressed in
black clothes of an old-looking cut—a Quaker, I believe.
Never shall I forget an apple-tart he invited me to try
as a proof of the merit of the "Wellington" apple.   It
was not only good, but beautiful; the cooked apple
looking rosy and transparent, and most inviting.   He
told me he was an ardent preacher of total abstinence,
and took me to a grassy, shady place among the nuts,
where there was an upright stone slab, like a tomb-
stone, with the inscription :

TO ALCOHOL.

He had dug a grave, and poured into it a quantity of
wine and beer and spirits, and placed the stone as a
memorial of his abhorrence of drink.   The whole thing
remains in my mind like a picture—the shady groves
of old nuts, in tenderest early leaf, the pale Daffodils,
the mighty chained mastiffs with bloodshot eyes and
murderous fangs, the brawny, wholesome forewoman,

and the trim old gentleman in black. It was the only
nursery I ever saw where one would expect to see
fairies on a summer's night.

I never tire of admiring and praising *Iris stylosa*,
which has proved itself such a good plant for English
gardens; at any rate, for those in our southern counties. Lovely in form and colour, sweetly-scented and
with admirable foliage, it has in addition to these
merits the unusual one of a blooming season of six
months' duration. The first flowers come with the
earliest days of November, and its season ends with a
rush of bloom in the first half of April. Then is the
time to take up old tufts and part them, and plant
afresh; the old roots will have dried up into brown
wires, and the new will be pushing. It thrives in
rather poor soil, and seems to bloom all the better for
having its root-run invaded by some stronger plant.
When I first planted a quantity I had brought from its
native place, I made the mistake of putting it in a
well-prepared border. At first I was delighted to see
how well it flourished, but as it gave me only thick
masses of leaves a yard long, and no flowers, it was
clear that it wanted to be less well fed. After changing it to poor soil, at the foot of a sunny wall close to
a strong clump of Alströmeria, I was rewarded with a
good crop of flowers; and the more the Alströmeria
grew into it on one side and *Plumbago Larpenti* on the
other, the more freely the brave little Iris flowered.
The flower has no true stem; what serves as a stem,

sometimes a foot long, is the elongated style, so that the seed-pod has to be looked for deep down at the base of the tufts of leaves, and almost under ground. The specific name, *stylosa*, is so clearly descriptive, that one regrets that the longer, and certainly uglier, *unguicularis* should be preferred by botanists.

What a delight it was to see it for the first time in its home in the hilly wastes, a mile or two inland from the town of Algiers! Another lovely blue Iris was there too, *I. alata* or *scorpioides*, growing under exactly the same conditions; but this is a plant unwilling to be acclimatised in England. What a paradise it was for flower-rambles, among the giant Fennels and the tiny orange Marigolds, and the immense bulbs of *Scilla maritima* standing almost out of the ground, and the many lovely Bee-orchises and the fairy-like *Narcissus serotinus*, and the groves of Prickly Pear wreathed and festooned with the graceful tufts of bell-shaped flower and polished leaves of *Clematis cirrhosa!*

It was in the days when there were only a few English residents, but among them was the Rev. Edwyn Arkwright, who by his happy discovery of a white-flowered *Iris stylosa*, the only one that has been found wild, has enriched our gardens with a most lovely variety of this excellent plant. I am glad to be able to quote his own words :—

"The finding of the white *Iris stylosa* belongs to the happy old times twenty-five years ago, when there

were no social duties and no vineyards[1] in Algiers.
My two sisters and I bought three horses, and rode wild
every day in the scrub of Myrtle, Cistus, Dwarf Oak,
&c. It was about five miles from the town, on what
is called the 'Sahel,' that the one plant grew that I
was told botanists knew ought to exist, but with all
their searching had never found. I am thankful that
I dug it up instead of picking it, only knowing that it
was a pretty flower. Then after a year or two Durando
saw it, and took off his hat to it, and told me what a
treasure it was, and proceeded to send off little bits to
his friends; and among them all, Ware of Tottenham
managed to be beforehand, and took a first-class certi-
ficate for it. It is odd that there should never have
been another plant found, for there never was such a
free-growing and multiplying plant. My sister in
Herefordshire has had over fifty blooms this winter;
but we count it by thousands, and it is *the* feature in
all decorations in every English house in Algiers."

Throughout January, and indeed from the middle
of December, is the time when outdoor flowers for
cutting and house decoration are most scarce; and yet
there are Christmas Roses and yellow Jasmine and
Laurustinus, and in all open weather *Iris stylosa* and
Czar Violets. A very few flowers can be made to look
well if cleverly arranged with plenty of good foliage;
and even when a hard and long frost spoils the few

---

[1] The planting of large vineyards, in some cases of private enterprise, had not proved a financial success.

blooms that would otherwise be available, leafy branches alone are beautiful in rooms. But, as in all matters that have to do with decoration, everything depends on a right choice of material and the exercise of taste in disposing it. Red-tinted Berberis always looks well alone, if three or four branches are boldly cut from two to three feet long. Branches of the spotted Aucuba do very well by themselves, and are specially beautiful in blue china; the larger the leaves and the bolder the markings, the better. Where there is an old Exmouth Magnolia that can spare some small branches, nothing makes a nobler room-ornament. The long arching sprays of Alexandrian Laurel do well with green or variegated Box, and will live in a room for several weeks. Among useful winter leaves of smaller growth, those of *Epimedium pinnatum* have a fine red colour and delicate veining, and I find them very useful for grouping with greenhouse flowers of delicate texture. *Gaultheria shallon* is at its best in winter, and gives valuable branches and twigs for cutting; and much to be prized are sprays of the Japan Privet, with its tough, highly-polished leaves, so much like those of the orange. There is a variegated Eurybia, small branches of which are excellent; and always useful are the gold and silver Hollies.

There is a little plant, *Ophiopogon spicatum*, that I grow in rather large quantity for winter cutting, the leaves being at their best in the winter months. They are sword-shaped and of a lively green colour, and are arranged in flat sheaves after the manner of a flag-Iris. I pull up a whole plant at a time—a two-year-old plant is a spreading tuft of the little sheaves—and wash it and cut away the groups of leaves just at the root, so that they are held together by the root-stock. They last long in water, and are beautiful with Roman Hyacinths or Freesias or *Iris stylosa* and many other flowers. The leaves of Megaseas, especially those of the *cordifolia* section, colour grandly in winter, and look fine in a large bowl with the largest blooms of Christmas Roses, or with forced Hyacinths. Much useful material can be found among Ivies, both of the wild and garden kinds. When they are well established they generally throw out rather woody front shoots; these are the ones to look out for, as they stand out with a certain degree of stiffness that makes them easier to arrange than weaker trailing pieces.

I do not much care for dried flowers—the bulrush and pampas-grass decoration has been so much over-done, that it has become wearisome—but I make an exception in favour of the flower of *Eulalia japonica*, and always give it a place. It does not come to its full beauty out of doors; it only finishes its growth late in October, and therefore does not have time to dry and expand. I grew it for many years before finding out that the closed and rather draggled-looking heads would open perfectly in a warm room. The uppermost leaf often confines the flower, and should be taken off

B

to release it; the flower does not seem to mature quite enough to come free of itself. Bold masses of Helichrysum certainly give some brightness to a room during the darkest weeks of winter, though the brightest yellow is the only one I much care to have; there is a look of faded tinsel about the other colourings. I much prize large bunches of the native Iris berries, and grow it largely for winter room-ornament.

Among the many valuable suggestions in Mrs. Earle's delightful book, "Pot-pourri from a Surrey Garden," is the use indoors of the smaller coloured gourds. As used by her they give a bright and cheerful look to a room that even flowers can not surpass.

A WILD JUNIPER.

CHAPTER III

FEBRUARY

Distant promise of summer—Ivy-berries—Coloured leaves—*Berberis Aquifolium*—Its many merits—Thinning and pruning shrubs—Lilacs—Removing suckers—Training *Clematis flammula*—Forms of trees—Juniper, a neglected native evergreen—Effect of snow—Power of recovery—Beauty of colour—Moss-grown stems.

THERE is always in February some one day, at least, when one smells the yet distant, but surely coming, summer. Perhaps it is a warm, mossy scent that greets one when passing along the southern side of a hedge-bank; or it may be in some woodland opening, where the sun has coaxed out the pungent smell of the trailing ground Ivy, whose blue flowers will soon appear; but the day always comes, and with it the glad certainty that summer is nearing, and that the good things promised will never fail.

How strangely little of positive green colour is to be seen in copse and woodland. Only the moss is really green. The next greenest thing is the northern sides of the trunks of beech and oak. Walking southward they are all green, but looking back they are silver-grey. The undergrowth is of brambles and sparse fronds of withered bracken; the bracken less beaten down than usual, for the winter has been without snow; only where the soil is deeper, and the fern has grown more tall and rank, it has fallen into thick, almost felted masses, and the stalks all lying one way make the heaps look like lumps of fallen thatch. The bramble leaves—last year's leaves, which are held all the winter—are of a dark, blackish-bronze colour, or nearly red where they have seen the sun. Age seems to give them a sort of hard surface and enough of a polish to reflect the sky; the young leaves that will come next month are almost woolly at first. Grassy tufts show only bleached bents, so tightly matted, that one wonders how the delicate young blades will be able to spear through. Ivy-berries, hanging in thick clusters, are still in beauty; they are so heavy that they weigh down the branches. There is a peculiar beauty in the form and veining of the plain-shaped leaves belonging to the mature or flowering state that the plant reaches when it can no longer climb, whether on a wall six feet high or on the battlements of a castle. Cuttings grown from such portions retain this habit, and form densely-flowering bushes of compact shape.

Beautiful colouring is now to be seen in many of the plants whose leaves do not die down in winter. Foremost amongst these is the Foam-flower (*Tiarella cordifolia*), whose leaves, now lying on the ground, show bright colouring, inclining to scarlet, crimson,

and orange. *Tellima*, its near relation, is also well
coloured. *Galax aphylla*, with its polished leaves of
hard texture, and stalks almost as stiff as wire, is
nearly as bright; and many of the Megaseas are of a
fine bronze red, the ones that colour best being the
varieties of the well-known *M. crassifolia* and *M. cordifolia*. Among shrubs, some of the nearly allied genera,
popularly classed under the name Andromeda, are
beautiful in reddish colour passing into green, in some
of the leaves by tender gradation, and in others by
bold splashing. *Berberis Aquifolium* begins to colour
after the first frosts; though some plants remain
green, the greater number take on some rich tinting
of red or purple, and occasionally in poor soil and in
full sun a bright red that may almost be called scarlet.

What a precious thing this fine old Berberis is!
What should we do in winter without its vigorous
masses of grand foliage in garden and shrubbery, to
say nothing of its use indoors? Frequent as it is in
gardens, it is seldom used as well or thoughtfully as
it deserves. There are many places where, between
garden and wood, a well-considered planting of Berberis, combined with two or three other things of
larger stature, such as the fruiting Barberry, and White-
thorn and Holly, would make a very enjoyable piece
of shrub wild-gardening. When one reflects that
Berberis Aquifolium is individually one of the handsomest of small shrubs, that it is at its very best in
mid-winter, that every leaf is a marvel of beautiful

drawing and construction, and that its ruddy winter colouring is a joy to see, enhanced as it is by the glistening brightness of the leaf-surface; and further, when one remembers that in spring the whole picture changes—that the polished leaves are green again, and the bushes are full of tufted masses of brightest yellow bloom, and fuller of bee-music than any other plant then in flower; and that even then it has another season of beauty yet to come, when in the days of middle summer it is heavily loaded with the thick-clustered masses of berries, covered with a brighter and bluer bloom than almost any other fruit can show,—when one thinks of all this brought together in one plant, it seems but right that we should spare no pains to use it well. It is the only hardy shrub I can think of that is in one or other of its varied forms of beauty throughout the year. It is never leafless or untidy; it never looks mangy like an Ilex in April, or moulting like a Holly in May, or patchy and unfinished like Yew and Box and many other ever-greens when their young leafy shoots are sprouting.

We have been thinning the shrubs in one of the rather large clumps next to the lawn, taking the older wood in each clump right out from the bottom and letting more light and air into the middle. Weigelas grow fast and very thick. Quite two-thirds have been cut out of each bush of Weigela, Philadelphus, and Ribes, and a good bit out of Ceanothus, "Gloire de Versailles," my favourite of its kind, and all the oldest

wood from *Viburnum plicatus*. The stuff cut out
makes quite a respectable lot of faggoting. How
extremely dense and hard is the wood of Philadelphus! as close-grained as Box, and almost as hard
as the bright yellow wood of Berberis.

Some of the Lilacs have a good many suckers from
the root, as well as on the lower part of the stem.
These must all come away, and then the trees will
have a good dressing of manure. They are greedy
feeders, and want it badly in our light soil, and surely
no flowering shrub more truly deserves it. The Lilacs
I have are some of the beautiful kinds raised in
France, for which we can never be thankful enough
to our good neighbours across the Channel. The
white variety, "Marie Legraye," always remains my
favourite. Some are larger and whiter, and have
the trusses more evenly and closely filled, but this
beautiful Marie fills one with a satisfying conviction
as of something that is just right, that has arrived
at the point of just the best and most lovable kind
of beauty, and has been wisely content to stay there,
not attempting to pass beyond and excel itself. Its
beauty is modest and reserved, and temperate and
full of refinement. The colour has a deliciously-
tender warmth of white, and as the truss is not
over-full, there is room for a delicate play of warm
half-light within its recesses. Among the many
beautiful coloured Lilacs, I am fond of Lucie Baltet
and Princesse Marie. There may be better flowers

from the ordinary florist point of view, but these
have the charm that is a good garden flower's
most precious quality.   I do not like the cold, heavy-
coloured ones of the bluish-slaty kinds.   No shrub
is hardier than the Lilac; I believe they flourish
even within the Arctic Circle.   It is very nearly allied
to Privet; so nearly, that the oval-leaved Privet is
commonly used as a stock.   Standard trees flower
much better than bushes; in this form all the strength
seems to go directly to the flowering boughs.   No
shrub is more persistent in throwing up suckers
from the root and from the lower part of the stem,
but in bush trees as well as in standards they should
be carefully removed every year.   In the case of
bushes, three or four main stems will be enough to
leave.   When taking away suckers of any kind whatever, it is much better to tear them out than to
cut them off.   A cut, however close, leaves a base
from which they may always spring again, but if
pulled or wrenched out they bring away with them
the swollen base that, if left in, would be a likely source
of future trouble.

Before the end of February we must be sure to
prune and train any plants there may be of *Clematis
flammula*.   Its growth is so rapid when once it begins,
that if it is overlooked it soon grows into a tangled
mass of succulent weak young stuff, quite unmanage-
able two months hence, when it will be hanging about
in helpless masses, dead and living together.   If it

is left till then, one can only engirdle the whole thing
with a soft tarred rope and sling it up somehow or
anyhow.   But if taken now, when the young growths
are just showing at the joints, the last year's mass can
be untangled, the dead and the over-much cut out,
and the best pieces trained in.   In gardening, the
interests of the moment are so engrossing that one
is often tempted to forget the future; but it is well
to remember that this lovely and tenderly-scented
Clematis will be one of the chief beauties of September,
and well deserves a little timely care.

In summer-time one never really knows how beautiful are the forms of the deciduous trees.   It is only in
winter, when they are bare of leaves, that one can fully
enjoy their splendid structure and design, their admirable qualities of duly apportioned strength and grace of
poise, and the way the spread of the many-branched
head has its equivalent in the wide-reaching ground-
grasp of the root.   And it is interesting to see how,
in the many different kinds of tree, the same laws are
always in force, and the same results occur, and yet by
the employment of what varied means.   For nothing
in the growth of trees can be much more unlike than
the habit of the oak and that of the weeping willow,
though the unlikeness only comes from the different
adjustment of the same sources of power and the same
weights, just as in the movement of wind-blown leaves
some flutter and some undulate, while others turn over
and back again.   Old apple-trees are specially notice-

able for their beauty in winter, when their extremely graceful shape, less visible when in loveliness of spring bloom or in rich bounty of autumn fruit, is seen to fullest advantage.

Few in number are our native evergreens, and for that reason all the more precious.   One of them, the common Juniper, is one of the best of shrubs either for garden or wild ground, and yet, strangely enough, it is so little appreciated that it is scarcely to be had in nurseries.   Chinese Junipers, North American Junipers, Junipers from Spain and Greece, from Nepaul and Siberia, may be had, but the best Juniper of all is very rarely grown.   Were it a common tree one could see a sort of reason (to some minds) for over-looking it, but though it is fairly abundant on a few hill-sides in the southern counties, it is by no means widely distributed throughout the country.   Even this reason would not be consistent with common practice, for the Holly is abundant throughout England, and yet is to be had by the thousand in every nursery.   Be the reason what it may, the common Juniper is one of the most desirable of evergreens, and is most un-deservedly neglected.   Even our botanists fail to do it justice, for Bentham describes it as a low shrub growing two feet, three feet, or four feet high.   I quote from memory only; these may not be the words, but this is the sense of his description.   He had evidently seen it on the chalk downs only, where such a portrait of it is exactly right.   But in our sheltered uplands, in

SCOTCH FIRS THROWN ON TO FROZEN WATER BY SNOWSTORM.

sandy soil, it is a small tree of noble aspect, twelve
to twenty-eight feet high. In form it is extremely
variable, for sometimes it shoots up on a single stem
and looks like an Italian Cypress or like the upright
Chinese Juniper, while at other times it will have two
or more tall spires and a dense surrounding mass of
lower growth, while in other cases it will be like a
quantity of young trees growing close together, and
yet the trees in all these varied forms may be nearly
of an age.

The action of snow is the reason of this unlikeness
of habit. If, when young, the tree happens to have
one main stem strong enough to shoot up alone, and
if at the same time there come a sequence of winters
without much snow, there will be the tall, straight,
cypress-like tree. But if, as is more commonly the
case, the growth is divided into a number of stems of
nearly equal size, sooner or later they are sure to be
laid down by snow. Such a winter storm as that of
the end of December 1886 was especially disastrous to
Junipers. Snow came on early in the evening in this
district, when the thermometer was barely at freezing point and there was no wind. It hung on the
trees in clogging masses, with a lowering temperature
that was soon below freezing. The snow still falling
loaded them more and more ; then came the fatal
wind, and all through that night we heard the breaking trees. When morning came there were eighteen
inches of snow on the ground, and all the trees that

could be seen, mostly Scotch fir, seemed to be completely wrecked. Some were entirely stripped of
branches, and stood up bare, like scaffold-poles. Until
the snow was gone or half gone, no idea could be
formed of the amount of damage done to shrubs; all
were borne down and buried under the white rounded
masses. A great Holly on the edge of the lawn, nearly
thirty feet high and as much in spread, whose head in
summer is crowned with a great tangle of Honeysuckle,
had that crowned head lying on the ground weighted
down by the frozen mass. But when the snow was
gone and all the damage could be seen, the Junipers
looked worse than anything. What had lately been
shapely groups were lying perfectly flat, the bare-
stemmed, leafless portions of the inner part of the
group showing, and looking like a faggot of dry brush-
wood, that, having been stood upright, had burst its
band and fallen apart in all directions. Some, whose
stems had weathered many snowy winters, now had
them broken short off half-way up; while others escaped
with bare life, but with the thick, strong stem broken
down, the heavy head lying on the ground, and the
stem wrenched open at the break, like a half-untwisted
rope. The great wild Junipers were the pride of our
stretch of heathy waste just beyond the garden, and
the scene of desolation was truly piteous, for though
many of them already bore the marks of former
accidents, never within our memory had there been
such complete and comprehensive destruction.

JUNIPER, LATELY WRECKED BY SNOWSTORM.

OLD JUNIPER, SHOWING FORMER INJURIES.

But now, ten years later, so great is their power of recovery, that there are the same Junipers, and, except in the case of those actually broken off, looking as well as ever. For those with many stems that were laid down flat have risen at the tips, and each tip looks like a vigorous young ten-year-old tree. What was formerly a massive, bushy-shaped Juniper, some twelve feet to fifteen feet high, now covers a space thirty feet across, and looks like a thick group of closely-planted, healthy young ones. The half broken-down trees have also risen at the tips, and are full of renewed vigour. Indeed, this breaking down and splitting open seems to give them a new energy, for individual trees that I have known well, and observed to look old and over-worn, and to all appearance on the downward road of life, after being broken and laid down by snow, have, some years later, shot up again with every evidence of vigorous young life. It would be more easily accounted for if the branch rooted where it touched the ground, as so many trees and bushes will do ; but as far as I have been able to observe, the Juniper does not " layer " itself. I have often thought I had found a fine young one fit for transplanting, but on clearing away the moss and fern at the supposed root have found that it was only the tip of a laid-down branch of a tree perhaps twelve feet away. In the case of one of our trees, among a group of laid-down and grown-up branches, one old central trunk has sur-vived. It is now so thick and strong, and has so

little top, that it will be likely to stand till it falls
from sheer old age. Close to it is another, whose
main stem was broken down about five feet from the
ground; now, what was the head rests on the earth
nine feet away, and a circle of its outspread branches
has become a wholesome group of young upright
growths, while at the place where the stem broke, the
half-opened wrench still shows as clearly as on the
day it was done.

Among the many merits of the Juniper, its tenderly
mysterious beauty of colouring is by no means the
least; a colouring as delicately subtle in its own way
as that of cloud or mist, or haze in warm, wet woodland. It has very little of positive green; a suspicion
of warm colour in the shadowy hollows, and a blue-
grey bloom of the tenderest quality imaginable on the
outer masses of foliage. Each tiny, blade-like leaf has
a band of dead, palest bluish-green colour on the
upper surface, edged with a narrow line of dark green
slightly polished; the back of the leaf is of the same
full, rather dark green, with slight polish; it looks as if
the green back had been brought up over the edge of
the leaf to make the dark edging on the upper surface.
The stems of the twigs are of a warm, almost foxy
colour, becoming darker and redder in the branches.
The tips of the twigs curl over or hang out on all sides
towards the light, and the "set" of the individual twigs
is full of variety. This arrangement of mixed colouring and texture, and infinitely various positions of the

spiny little leaves, allows the eye to penetrate unconsciously a little way into the mass, so that one sees as
much tender shadow as actual leaf-surface, and this is
probably the cause of the wonderfully delicate and, so
to speak, intangible quality of colouring. Then, again,
where there is a hollow place in a bush, or group, showing
a cluster of half-dead stems, at first one cannot tell
what the colour is, till with half-shut eyes one becomes
aware of a dusky and yet luminous purple-grey.

The merits of the Juniper are not yet done with,
for throughout the winter (the time of growth of moss
and lichen) the rugged-barked old stems are clothed
with loveliest pale-green growths of a silvery quality.
Standing before it, and trying to put the colour into
words, one repeats, again and again, pale-green silver—
palest silvery green! Where the lichen is old and
dead it is greyer; every now and then there is a touch
of the orange kind, and a little of the branched stag-
horn pattern so common on the heathy ground. Here
and there, as the trunk or branch is increasing in
girth, the silvery, lichen-clad, rough outer bark has
parted, and shows the smooth, dark-red inner bark;
the outer covering still clinging over the opening, and
looking like grey ribands slightly interlaced. Many
another kind of tree-stem is beautiful in its winter
dress, but it is difficult to find any so full of varied
beauty and interest as that of the Juniper; it is one
of the yearly feasts that never fails to delight and
satisfy.

CHAPTER IV

MARCH

Flowering bulbs — Dog-tooth Violet — Rock-garden — Variety of Rhododendron foliage — A beautiful old kind — Suckers on grafted plants—Plants for filling up the beds—Heaths—Andromedas — Lady Fern — *Lilium auratum* — Pruning Roses— Training and tying climbing plants—Climbing and free-growing Roses—-The Vine the best wall-covering—Other climbers— Wild Clematis—Wild Rose.

In early March many and lovely are the flowering bulbs, and among them a wealth of blue, the more precious that it is the colour least frequent among flowers. The blue of *Scilla sibirica*, like all blues that have in them a suspicion of green, has a curiously penetrating quality; the blue of *Scilla bifolia* does not attack the eye so smartly. *Chionodoxa sardensis* is of a full and satisfying colour, that is enhanced by the small space of clear white throat. A bed of it shows very little variation in colour. *Chionodoxa Lucilliæ*, on the other hand, varies greatly; one may pick out light and dark blue, and light and dark of almost lilac colour. The variety *C. gigantea* is a fine plant. There are some pretty kinds of *Scilla bifolia* that were raised by the Rev. J. G. Nelson of Aldborough, among them a tender

flesh-colour and a good pink. *Leucojum vernum*, with
its clear white flowers and polished dark-green leaves,
is one of the gems of early March; and, flowering at
the same time, no flower of the whole year can show a
more splendid and sumptuous colour than the purple
of *Iris reticulata*. Varieties have been raised, some
larger, some nearer blue, and some reddish purple, but
the type remains the best garden flower. *Iris stylosa*,
in sheltered nooks open to the sun, when well established, gives flower from November till April, the
strongest rush of bloom being about the third week in
March. It is a precious plant in our southern counties,
delicately scented, of a tender and yet full lilac-blue.
The long ribbon-like leaves make handsome tufts, and
the sheltered place it needs in our climate saves the
flowers from the injury they receive on their native
windy Algerian hills, where they are nearly always torn
into tatters.

What a charm there is about the common Dog-
tooth Violet; it is pretty everywhere, in borders, in the
rock-garden, in all sorts of corners. But where it looks
best with me is in a grassy place strewn with dead
leaves, under young oaks, where the garden joins the
copse. This is a part of the pleasure-ground that has
been treated with some care, and has rewarded thought
and labour with some success, so that it looks less as if
it had been planned than as if it might have come
naturally. At one point the lawn, trending gently upward, runs by grass paths into a rock-garden, planted

c

mainly with dwarf shrubs. Here are Andromedas,
Pernettyas, Gaultherias, and Alpine Rhododendron,
and with them three favourites whose crushed leaves
give a grateful fragrance, Sweet Gale, *Ledum palustre*,
and *Rhododendron myrtifolium*. The rock part is unobtrusive; where the ground rises rather quickly are a
couple of ridges made of large, long lumps of sandstone, half buried, and so laid as to give a look of
natural stratification. Hardy Ferns are grateful for
the coolness of their northern flanks, and Cyclamens
are happy on the ledges. Beyond and above is the
copse, or thin wood of young silver Birch and Holly,
in summer clothed below with bracken, but now brist-
ling with the bluish spears of Daffodils and buds
that will soon burst into bloom. The early Pyrenean
Daffodil is already out, gleaming through the low-
toned copse like lamps of pale yellow light. Where
the rough path enters the birch copse is a cheerfully
twinkling throng of the Dwarf Daffodil (*N. nanus*),
looking quite at its best on its carpet of moss and fine
grass and dead leaves. The light wind gives it a
graceful, dancing movement, with an active spring
about the upper part of the stalk. Some of the heavier
trumpets not far off answer to the same wind with
only a ponderous, leaden sort of movement.

Farther along the garden joins the wood by a
plantation of Rhododendrons and broad grassy paths,
and farther still by a thicket of the free-growing Roses,
some forming fountain-like clumps nine paces in dia-

meter, and then again by masses of flowering shrubs, gradating by means of Sweetbriar, Water-elder, Dogwood, Medlar, and Thorn from garden to wild wood.

Now that the Rhododendrons, planted nine years ago, have grown to a state and size of young maturity, it is interesting to observe how much they vary in foliage, and how clearly the leaves show their relative degrees of relationship to their original parents, the wild mountain plants of Asia Minor and the United States. These, being two of the hardiest kinds, were the ones first chosen by hybridisers, and to these kinds we owe nearly all of the large numbers of beautiful garden Rhododendrons now in cultivation. The ones more nearly related to the wild *R. ponticum* have long, narrow, shining dark-green leaves, while the varieties that incline more to the American *R. catawbiense* have the leaves twice as broad, and almost rounded at the shoulder where they join the stalk; moreover, the surface of the leaf has a different texture, less polished, and showing a grain like morocco leather. The colour also is a lighter and more yellowish green, and the bush is not so densely branched. The leaves of all the kinds are inclined to hang down in cold weather, and this habit is more clearly marked in the *catawbiense* varieties.

There is one old kind called *Multum maculatum*— I dare say one of the earliest hybrids—for which I have a special liking. It is now despised by florists, because the flower is thin in texture and the petal narrow, and the truss not tightly filled. Nevertheless I find it quite the most beautiful Rhododendron as a cut flower, perhaps just because of these unorthodox qualities. And much as I admire the great bouncing beauties that are most justly the pride of their raisers, I hold that this most refined and delicate class of beauty equally deserves faithful championship. The flowers of this pretty old kind are of a delicate milk-white, and the lower petals are generously spotted with a rosy-scarlet of the loveliest quality. The leaves are the longest and narrowest and darkest green of any kind I know, making the bush conspicuously hand-some in winter. I have to confess that it is a shy bloomer, and that it seems unwilling to flower in a young state, but I think of it as a thing so beautiful and desirable as to be worth waiting for.

Within March, and before the busier season comes upon us, it is well to look out for the suckers that are likely to come on grafted plants. They may generally be detected by the typical *ponticum* leaf, but if the foliage of a branch should be suspicious and yet doubt-ful, if on following the shoot down it is seen to come straight from the root and to have a redder bark than the rest, it may safely be taken for a robber. Of course the invading stock may be easily seen when in flower, but the good gardener takes it away before it has this chance of reproaching him. A lady visitor last year told me with some pride that she had a most wonder-ful Rhododendron in bloom; all the flower in the middle was crimson, with a ring of purple-flowered branches outside. I am afraid she was disappointed when I offered condolence instead of congratulation, and had to tell her that the phenomenon was not uncommon among neglected bushes.

When my Rhododendron beds were first planted, I followed the usual practice of filling the outer empty spaces of the clumps with hardy Heaths. Perhaps it is still the best or one of the best ways to begin when the bushes are quite young; for if planted the right distance apart—seven to nine feet—there must be large bare spaces between; but now that they have filled the greater part of the beds, I find that the other plants I tried are more to my liking. These are, foremost of all, *Andromeda Catesbæi*, then Lady Fern, and then the dwarf *Rhododendron myrtifolium*. The main spaces between the young bushes I plant with *Cistus laurifolius*, a perfectly hardy kind; this grows much faster than the Rhododendrons, and soon fills the middle spaces; by the time that the best of its life is over—for it is a short-lived bush—the Rhododendrons will be wanting all the space. Here and there in the inner spaces I put groups of *Lilium auratum*, a Lily that thrives in a peaty bed, and that looks its best when growing through other plants; moreover, when the Rhododendrons are out of flower, the Lily, whose blooming season is throughout the late summer and autumn, gives a new beauty and interest to that part of the garden.

The time has come for pruning Roses, and for tying up and training the plants that clothe wall and fence and pergola. And this sets one thinking about climbing and rambling plants, and all their various ways and wants, and of how best to use them. One of my boundaries to a road is a fence about nine feet high, wall below and close oak paling above. It is planted with free-growing Roses of several types—Aimée Vibert, Madame Alfred Carrière, Reine Olga de Wurtemburg, and Bouquet d'Or, the strongest of the Dijon teas. Then comes a space of *Clematis Montana* and *Clematis flammula*, and then more Roses—Madame Plantier, Emélie Plantier (a delightful Rose to cut), and some of the grand Sweetbriars raised by Lord Penzance.

From midsummer onwards these Roses are continually cut for flower, and yield an abundance of quite the most ornamental class of bloom. For I like to have cut Roses arranged in a large, free way, with whole branches three feet or four feet long, easy to have from these free-growing kinds, that throw out branches fifteen feet long in one season, even on our poor, sandy soil, that contains no particle of that rich loam that Roses love. I think this same Reine Olga, the grand grower from which have come our longest and largest prunings, must be quite the best evergreen Rose, for it holds its full clothing of handsome dark-green leaves right through the winter. It seems to like hard pruning. I have one on a part of the pergola, but have no pleasure from it, as it has rushed

GARDEN DOOR-WAY WREATHED WITH CLEMATIS GRAVEOLENS.

COTTAGE PORCH WREATHED WITH THE DOUBLE
WHITE ROSE (*R. alba*).

up to the top, and nothing shows but a few naked stems.

One has to find out how to use all these different Roses. How often one sees the wrong Roses used as climbers on the walls of a house. I have seen a Gloire de Dijon covering the side of a house with a profitless reticulation of bare stem, and a few leaves and flowers looking into the gutter just under the edge of the roof. What are generally recommended as climbing Roses are too ready to ramp away, leaving bare, leggy growth where wall-clothing is desired. One of the best is climbing Aimée Vibert, for with very little pruning it keeps well furnished nearly to the ground, and with its graceful clusters of white bloom and healthy-looking, polished leaves is always one of the prettiest of Roses. Its only fault is that it does not shed its dead petals, but retains the whole bloom in dead brown clusters.

But if a Rose wishes to climb, it should be accommodated with a suitable place. That excellent old Rose, the Dundee Rambler, or the still prettier Garland Rose, will find a way up a Holly-tree, and fling out its long wreaths of tenderly-tinted bloom; and there can be no better way of using the lovely Himalayan *R. Brunonis*, with its long, almost blue leaves and wealth of milk-white flower. A common Sweetbriar will also push up among the branches of some dark evergreen, Yew or Holly, and throw out aloft its scented branches and rosy bloom, and look its very best.

But some of these same free Roses are best of all if left in a clear space to grow exactly as they will without any kind of support or training. So placed, they grow into large rounded groups. Every year, just after the young laterals on the last year's branches have flowered, they throw out vigorous young rods that arch over as they complete their growth, and will be the flower-bearers of the year to come.

Two kinds of Roses of rambling growth that are rather tender, but indispensable for beauty, are Fortune's Yellow and the Banksias. Pruning the free Roses is always rough work for the hands and clothes, but of all Roses I know, the worst to handle is Fortune's Yellow. The prickles are hooked back in a way that no care or ingenuity can escape; and whether it is their shape and power of cruel grip, or whether they have anything of a poisonous quality, I do not know; but whereas hands scratched and torn by Roses in general heal quickly, the wounds made by Fortune's Yellow are much more painful and much slower to get well. I knew an old labourer who died of a rose-prick. He used to work about the roads, and at cleaning the ditches and mending the hedges. For some time I did not see him, and when I asked another old countryman, "What's gone o' Master Trussler?" the answer was, "He's dead—died of a canker-bush." The wild Dog-rose is still the "canker" in the speech of the old people, and a thorn or prickle is still a "bush." A Dog-rose prickle had gone deep into the old hedger's

hand—a " bush " more or less was nothing to him, but
the neglected little wound had become tainted with
some impurity, blood-poisoning had set in, and my
poor old friend had truly enough " died of a canker-
bush."

The flowering season of Fortune's Yellow is a very
short one, but it comes so early, and the flowers have
such incomparable beauty, and are so little like those
of any other Rose, that its value is quite without doubt.
Some of the Tea Roses approach it in its pink and
copper colouring, but the loose, open, rather flaunting
form of the flower, and the twisted set of the petals,
display the colour better than is possible in any of the
more regular-shaped Roses.   It is a good plan to grow
it through some other wall shrub, as it soon gets bare
below, and the early maturing flowering tips are glad
to be a little sheltered by the near neighbourhood of
other foliage.

I do not think that there is any other Rose that
has just the same rich, butter colour as the Yellow
Banksian, and this unusual colouring is the more
distinct because each little Rose in the cluster is nearly
evenly coloured all over, besides being in such dense
bunches.   The season of bloom is very short, but the
neat, polished foliage is always pleasant to see throughout the year.   The white kind and the larger white
are both lovely as to the individual bloom, but they
flower so much more shyly that the yellow is much
the better garden plant.

But the best of all climbing or rambling plants, whether for wall or arbour or pergola, is undoubtedly the Grape-Vine. Even when trimly pruned and trained for fruit-bearing on an outer wall it is an admirable picture of leafage and fruit-cluster; but to have it in fullest beauty it must ramp at will, for it is only when the fast-growing branches are thrown out far and wide that it fairly displays its graceful vigour and the generous magnificence of its incomparable foliage.

The hardy Chasselas, known in England by the rather misleading name Royal Muscadine, is one of the best, both for fruit and foliage. The leaves are of moderate size, with clearly serrated edges and that strongly waved outline that gives the impression of powerful build, and is, in fact, a mechanical contrivance intended to stiffen the structure. The colour of the leaves is a fresh, lively green, and in autumn they are prettily marbled with yellow. Where a very large-leaved Vine is wanted nothing is handsomer than the North American *Vitis Labrusca* or the Asiatic *Vitis Coignettii*, whose autumn leaves are gorgeously coloured. For a place that demands more delicate foliage there is the Parsley-Vine, that has a delightful look of refinement, and another that should not be forgotten is the Claret-Vine, with autumnal colouring of almost scarlet and purple, and abundance of tightly clustered black fruit, nearly blue with a heavy bloom.

Many an old house and garden can show the far-

WILD HOP, ENTWINING WORMWOOD AND COW-PARSNIP.

rambling power of the beautiful *Wistaria Chinensis*, and of the large-leaved *Aristolochia Sipho*, one of the best plants for covering a pergola, and of the varieties of *Ampelopsis*, near relations of the Grape-Vine. The limit of these notes only admits of mention of some of the more important climbers; but among these the ever-delightful white Jasmine must have a place. It will ramble far and fast if it has its own way, but then gives little flower; but by close winter pruning it can be kept full of bloom and leaf nearly to the ground.

The woods and hedges have also their beautiful climbing plants. Honeysuckle in suitable conditions will ramble to great heights—in this district most noticeable in tall Hollies and Junipers as well as in high hedges. The wild Clematis is most frequent on the chalk, where it laces together whole hedges and rushes up trees, clothing them in July with long wreaths of delicate bloom, and in September with still more conspicuous feathery seed. For rapid growth perhaps no English plant outstrips the Hop, growing afresh from the root every year, and almost equalling the Vine in beauty of leaf. The two kinds of wild Bryony are also herbaceous climbers of rapid growth, and among the most beautiful of our hedge plants.

The wild Roses run up to great heights in hedge and thicket, and never look so well as when among the tangles of mixed growth of wild forest land or clambering through some old gnarled thorn-tree. The common

Brambles are also best seen in these forest groups; these again in form of leaf show somewhat of a vine-like beauty.

In the end of March, or at any time during the month when the wind is in the east or north-east, all increase and development of vegetation appears to cease. As things are, so they remain. Plants that are in flower retain their bloom, but, as it were, under protest. A kind of sullen dulness pervades all plant life. Sweet-scented shrubs do not give off their fragrance; even the woodland moss and earth and dead leaves withhold their sweet, nutty scent. The surface of the earth has an arid, infertile look; a slight haze of an ugly grey takes the colour out of objects in middle distance, and seems to rob the flowers of theirs, or to put them out of harmony with all things around. But a day comes, or, perhaps, a warmer night, when the wind, now breathing gently from the south-west, puts new life into all growing things. A marvellous change is wrought in a few hours. A little warm rain has fallen, and plants, invisible before, and doubtless still underground, spring into glad life.

What an innocent charm there is about many of the true spring flowers. Primroses of many colours are now in bloom, but the prettiest, this year, is a patch of an early blooming white one, grouped with a delicate lilac. Then comes *Omphalodes verna*, with its flowers of brilliant blue and foliage of brightest green, better described by its pretty north-country name,

Blue-eyed Mary. There are Violets of many colours, but daintiest of all is the pale-blue St. Helena; whether it is the effect of its delicate colouring, or whether it has really a better scent than other varieties of the common Violet, I cannot say, but it always seems to have a more refined fragrance.

CHAPTER V

APRIL

IN early April there is quite a wealth of flower
among plants that belong half to wood and half to
garden. *Epimedium pinnatum*, with its delicate, orchid-
like spike of pale-yellow bloom, flowers with its last
year's leaves, but as soon as it is fully out the young
leaves rush up, as if hastening to accompany the
flowers. *Dentaria pinnata*, a woodland plant of Switzerland and Austria, is one of the handsomest of the
white-flowered *cruciferæ*, with well-filled heads of twelve
to fifteen flowers, and palmate leaves of freshest green.
Hard by, and the best possible plant to group with it,
is the lovely Virginian Cowslip (*Mertensia virginica*),
the very embodiment of the freshness of early spring.
The sheaf of young leafage comes almost black out

46

of the ground, but as the leaves develop, their dull,
lurid colouring changes to a full, pale green of a
curious texture, quite smooth, and yet absolutely un-
reflecting.   The dark colouring of the young leaves
now only remains as a faint tracery of veining on the
backs of the leaves and stalks, and at last dies quite
away as the bloom expands.   The flower is of a rare
and beautiful quality of colour, hard to describe—a
rainbow-flower of purple, indigo, full and pale blue,
and daintiest lilac, full of infinite variety and inde-
scribable charm.   The flowers are in terminal clusters,
richly filled ; lesser clusters springing from the axils
of the last few leaves and joining with the topmost
one to form a gracefully drooping head.   The lurid
colouring of the young leaves is recalled in the flower-
stems and calix, and enhances the colour effect of the
whole.   The flower of the common Dog-tooth Violet
is over, but the leaves have grown larger and handsomer.   They look as if, originally of a purplish-red
colour, some liquid had been dropped on them, making
confluent pools of pale green, lightest at the centre
of the drop.   The noblest plant of the same family
(*Erythronium giganteum*) is now in flower—a striking
and beautiful wood plant, with turn-cap shaped flowers
of palest straw-colour, almost white, and large leaves,
whose markings are not drop-like as in the more
familiar kind, but are arranged in a regular sequence
of bold splashings, reminding one of a *Maranta*.   The
flowers, single or in pairs, rise on stems a foot or fifteen

inches high; the throat is beautifully marked with
flames of rich bay on a yellow ground, and the handsome group of golden-anthered stamens and 'silvery
pistil make up a flower of singular beauty and refinement. That valuable Indian Primrose, *P. denticulata*,
is another fine plant for the cool edge or shady hollows
of woodland in rather good, deep soil.

But the glory of the copse just now consists in the
great stretches of Daffodils. Through the wood run
shallow, parallel hollows, the lowest part of each depression some nine paces apart. Local tradition says
they are the remains of old pack-horse roads; they
occur frequently in the forest-like heathery uplands
of our poor-soiled, sandy land, running, for the most
part, three or four together, almost evenly side by side.
The old people account for this by saying that when
one track became too much worn another was taken
by its side. Where these pass through the birch
copse the Daffodils have been planted in the shallow
hollows of the old ways, in spaces of some three yards
broad by thirty or forty yards long—one kind at a
time. Two of such tracks, planted with *Narcissus
princeps* and *N. Horsfieldi*, are now waving rivers of
bloom, in many lights and accidents of cloud and sunshine full of pictorial effect. The planting of Daffodils
in this part of the copse is much better than in any
other portions where there were no guiding track-ways,
and where they were planted in haphazard sprinklings.

The Grape Hyacinths are now in full bloom. It

DAFFODILS IN THE COPSE.

is well to avoid the common one (*Muscari racemosum*), at any rate in light soils, where it becomes a troublesome weed. One of the best is *M. conicum*; this, with the upright-leaved *M. botryoides*, and its white variety, are the best for general use, but the Plume Hyacinth, which flowers later, should have a place. *Ornithogalum nutans* is another of the bulbous plants that, though beautiful in flower, becomes so pestilent a weed that it is best excluded.

Where and how the early flowering bulbs had best be planted is a question of some difficulty. Perhaps the mixed border, where they are most usually put, is the worst place of all, for when in flower they only show as forlorn little patches of bloom rather far apart, and when their leaves die down, leaving their places looking empty, the ruthless spade or trowel stabs into them when it is desired to fill the space with some other plant. Moreover, when the border is manured and partly dug in the autumn, it is difficult to avoid digging up the bulbs just when they are in full root-growth. Probably the best plan is to devote a good space of cool bank to small bulbs and hardy ferns, planting the ferns in such groups as will leave good spaces for the bulbs; then as their leaves are going the fern fronds are developing and will cover the whole space. Another way is to have them among any groups of newly planted small shrubs, to be left there for spring blooming until the shrubs have covered their allotted space.

D

Many flowering shrubs are in beauty. *Andromeda floribunda* still holds its persistent bloom that has endured for nearly two months. The thick, drooping, tassel-like bunches of bloom of *Andromeda japonica* are just going over. *Magnolia stellata*, a compact bush some five feet high and wide, is white with the multitude of its starry flowers; individually they look half double, having fourteen to sixteen petals. *Forsythia suspensa*, with its graceful habit and tender yellow flower, is a much better shrub than *F. viridissima*, though, strangely enough, that is the one most commonly planted. Corchorus, with its bright-yellow balls, the fine old rosy Ribes, the Japan Quinces and their salmon-coloured relative *Pyrus Mauleii*, *Spiræa Thunbergi*, with its neat habit and myriads of tiny flowers, these make frequent points of beauty and interest.

In the rock-garden, *Cardamine trifoliata* and *Hutchinsia alpina* are conspicuous from their pure white flowers and neat habit; both have leaves of darkest green, as if the better to show off the bloom. *Ranunculus montanus* fringes the cool base of a large stone; its whole height not over three inches, though its bright-yellow flowers are larger than field buttercups. The surface of the petals is curiously brilliant, glistening and flashing like glass. *Corydalis capnoides* is a charming rock-plant, with flowers of palest sulphur colour, one of the neatest and most graceful of its family.

Magnolia stellata

DAFFODILS AMONG JUNIPERS WHERE GARDEN JOINS COPSE.

Border plants are pushing up vigorous green
growth; finest of all are the Veratrums, with their
bold, deeply-plaited leaves of brilliant green. Delphiniums and Oriental Poppies have also made strong
foliage, and Daylilies are conspicuous from their fresh
masses of pale greenery. Flag Iris have their leaves
three parts grown, and Pæonies are a foot or more
high, in all varieties of rich red colouring. It is a
good plan, when they are in beds or large groups, to
plant the dark-flowered Wallflowers among them, their
colour making a rich harmony with the reds of the
young Pæony growths.

There are balmy days in mid-April, when the
whole garden is fragrant with Sweetbriar. It is not
"fast of its smell," as Bacon says of the damask rose,
but gives it so lavishly that one cannot pass near a
plant without being aware of its gracious presence.
Passing upward through the copse, the warm air draws
a fragrance almost as sweet, but infinitely more subtle,
from the fresh green of the young birches; it is like
a distant whiff of Lily of the Valley. Higher still
the young leafage of the larches gives a delightful
perfume of the same kind. It seems as if it were the
office of these mountain trees, already nearest the
high heaven, to offer an incense of praise for their
new life.

Few plants will grow under Scotch fir, but a
notable exception is the Whortleberry, now a sheet of
brilliant green, and full of its arbutus-like, pink-tinged

flower. This plant also has a pleasant scent in the mass, difficult to localise, but coming in whiffs as it will.

The snowy Mespilus (*Amelancheir*) shows like puffs of smoke among the firs and birches, full of its milk-white, cherry-like bloom—a true woodland shrub or small tree. It loves to grow in a thicket of other trees, and to fling its graceful sprays about through their branches. It is a doubtful native, but naturalised and plentiful in the neighbouring woods. As seen in gardens, it is usually a neat little tree of shapely form, but it is more beautiful when growing at its own will in the high woods.

Marshy hollows in the valleys are brilliant with Marsh Marigold (*Caltha palustris*); damp meadows have them in plenty, but they are largest and hand-somest in the alder-swamps of our valley bottoms, where their great luscious clumps rise out of pools of black mud and water.

Adonis vernalis is one of the brightest flowers of the middle of April, the flowers looking large for the size of the plant. The bright-yellow, mostly eight-petalled, blooms are comfortably seated in dense, fennel-like masses of foliage. It makes strong tufts, that are the better for division every four years. The spring Bitter-vetch (*Orobus vernus*) blooms at the same time, a re-markably clean-looking plant, with its cheerful red and purple blossom and handsomely divided leaves. It is one of the toughest of plants to divide, the mass of

Hollyhock, Pink Beauty. (*See page 105.*)

Tiarella cordifolia.

black root is like so much wire. It is a good plan with plants that have such roots, when dividing-time comes, to take the clumps to a strong bench or block and cut them through at the crown with a sharp cold-chisel and hammer. Another of the showiest families of plants of the time is *Doronicum*. *D. Austriacum* is the earliest, but it is closely followed by the fine *D. Plantagineum.* The large form of wood Forget-me-not (*Myosotis sylvatica major*) is in sheets of bloom, opening pink and changing to a perfect blue. This is a great improvement on the old smaller one. Grouped with it, as an informal border, and in patches running through and among its clumps, is the Foam-flower (*Tiarella cordifolia*), whose flower in the mass looks like the wreaths of foam tossed aside by a mountain torrent. By the end of the month the Satin-leaf (*Heuchera Richardsoni*) is pushing up its richly-coloured leaves, of a strong bronze-red, gradating to bronze-green at the outer edge. The beauty of the plant is in the colour and texture of the foliage. To encourage full leaf growth the flower stems should be pinched out, and as they push up rather persistently, they should be looked over every few days for about a fortnight.

The Primrose garden is now in beauty, but I have so much to say about it that I have given it a chapter to itself towards the end of the book.

The Scotch firs are shedding their pollen; a flowering branch shaken or struck with a stick throws out a

pale-yellow cloud. Heavy rain will wash it out, so that
after a storm the sides of the roads and paths look as
if powdered sulphur had been washed up in drifts.
The sun has gained great power, and on still bright
days sharp *snicking* sounds are to be heard from the
firs. The dry cones of last year are opening, and the
flattened seeds with their paper-like edges are fluttering
down. Another sound, much like it but just a shade
sharper and more *staccato*, is heard from the Gorse
bushes, whose dry pods are flying open and letting fall
the hard, polished, little bean-like seeds.

Border Auriculas are making a brave show. Nothing
in the flower year is more interesting than a bed of good
seedlings of the Alpine class. I know nothing better
for pure beauty of varied colouring among early flowers.
Except in varieties of *Salpiglossis*, such rich gradation
of colour, from pale lilac to rich purple, and from rosy
pink to deepest crimson, is hardly to be found in any
one family of plants. There are varieties of cloudings
of smoky-grey, sometimes approaching black, invading,
and at the same time enhancing, the purer colours, and
numbers of shades of half-tones of red and purple, such
as are comprised within the term *murrey* of heraldry,
and tender blooms of one colour, sulphurs and milk-
whites—all with the admirable texture and excellent
perfume that belong to the "Bear's-ears" of old English gardens. For practical purposes the florist's definition of a good Auricula is of little value; that is for
the show-table, and, as Bacon says, "Nothing to the

TULIPA RETROFLEXA.

LATE SINGLE TULIPS, BREEDERS AND BYBLŒMEN.

true pleasure of a garden." The qualities to look for in the bed of seedlings are not the narrowing ones of proportion of eye to tube, of exact circle in the circumference of the individual pip, and so on, but to notice whether the plant has a handsome look and stands up well, and is a delightful and beautiful thing as a whole.

Tulips are the great garden flowers in the last week of April and earliest days of May. In this plant also the rule of the show-table is no sure guide to garden value; for the show Tulip, beautiful though it is, is of one class alone—namely, the best of the "broken" varieties of the self-coloured seedlings called "breeders." These seedlings, after some years of cultivation, change or "break" into a variation in which the original colouring is only retained in certain flames or feathers of colour, on a ground of either white or yellow. If the flames in each petal are symmetrical and well arranged, according to the rules laid down by the florist, it is a good flower; it receives a name, and commands a certain price. If, on the other hand, the markings are irregular, however beautiful the colouring, the flower is comparatively worthless, and is "thrown into mixture." The kinds that are the grandest in gardens are ignored by the florist. One of the best for graceful and delicate beauty is *Tulipa retroflexa*, of a soft lemon-yellow colour, and twisted and curled petals; then Silver Crown, a white flower with a delicate picotee-like thread of scarlet along the edge of the sharply pointed and

reflexed petals. A variety of this called Sulphur Crown
is only a little less beautiful. Then there is Golden
Crown, also with pointed petals and occasional thread-
ings of scarlet. Nothing is more gorgeous than the
noble *Gesneriana major*, with its great chalice of crim-
son-scarlet and pools of blue in the inner base of each
petal. The gorgeously flamed Parrot Tulips are indispensable, and the large double Yellow Rose, and
the early double white La Candeur. Of the later
kinds there are many of-splendid colouring and noble
port; conspicuous among them are *Reine d'Espagne*,
Couleur de vin, and *Bleu celeste*. There are beautiful
colourings of scarlet, crimson, yellow, chocolate, and
purple among the " breeders," as well as among the
so-called *bizarres* and *bybloemen* that comprise the show
kinds.

The best thing now in the rock-garden is a patch
of some twenty plants of *Arnebia echinoides*, always
happy in our poor, dry soil. It is of the Borage family,
a native of Armenia. It flowers in single or double-
branching spikes of closely-set flowers of a fine yellow.
Just below each indentation of the five-lobed corolla
is a spot which looks black by contrast, but is of a
very dark, rich, velvety brown. The day after the
flower has expanded the spot has faded to a moderate
brown, the next day to a faint tinge, and on the fourth
day it is gone. The legend, accounting for the spots,
says that Mahomet touched the flower with the tips of
his fingers, hence its English name of Prophet-flower.

The upper parts of the rock-garden that are beyond hand-reach are planted with dwarf shrubs, many of them sweetly scented either as to leaf or flower—*Gaultherias*, Sweet Gale, Alpine Rhododendron, *Skimmias*, *Pernettyas*, *Ledums*, and hardy Daphnes. *Daphne pontica* now gives off delicious wafts of fragrance, intensely sweet in the evening.

In March and April Daffodils are the great flowers for house decoration, coming directly after the Lent Hellebores. Many people think these beautiful late-flowering Hellebores useless for cutting because they live badly in water. But if properly prepared they live quite well, and will remain ten days in beauty. Directly they are cut, and immediately before putting in water, the stalks should be slit up three or four inches, or according to their length, and then put in deep, so that the water comes nearly up to the flowers; and so they should remain, in a cool place, for some hours, or for a whole night, after which they can be arranged for the room. Most of them are inclined to droop; it is the habit of the plant in growth; this may be corrected by arranging them with something stiff like Box or Berberis.

Anemone fulgens is a grand cutting flower, and looks well with its own leaves only or with flowering twigs of Laurustinus. Then there are Pansies, delightful things in a room, but they should be cut in whole branches of leafy stem and flower and bud. At first the growths are short and only suit dish-shaped things, but as the season goes on they grow longer and bolder, and graduate first into bowls and then into upright glasses. I think Pansies are always best without mixture of other flowers, and in separate colours, or only in such varied tints as make harmonies of one class of colour at a time.

The big yellow and white bunch Primroses are delightful room flowers, beautiful, and of sweetest scent. When full-grown the flower-stalks are ten inches long and more. Among the seedlings there are always a certain number that are worthless. These are pounced upon as soon as they show their bloom, and cut up for greenery to go with the cut flowers, leaving the root-stock with all its middle foliage, and cutting away the roots and any rough outside leaves.

When the first Daffodils are out and suitable greenery is not abundant in the garden (for it does not do to cut their own blades), I bring home hand-fuls of the wild Arum leaves, so common in roadside hedges, grasping the whole plant close to the ground; then a steady pull breaks it away from the tuber, and you have a fine long-stalked sheaf of leafage held together by its own underground stem. This should be prepared like the Lent Hellebores, by putting it deep in water for a time. I always think the trumpet Daffodils look better with this than with any other kind of foliage. When the wild Arum is full-grown the leaves are so large and handsome that they do quite well to ac-company the white Arum flowers from the greenhouse.

CHAPTER VI

MAY

WHILE May is still young, Cowslips are in beauty on the chalk lands a few miles distant, but yet within pleasant reach. They are finest of all in orchards, where the grass grows tall and strong under the half-shade of the old apple-trees, some of the later kinds being still loaded with bloom. The blooming of the Cowslip is the signal for a search for the Morell, one of the very best of the edible fungi. It grows in open woods or where the undergrowth has not yet grown high, and frequently in old parks and pastures near or under elms. It is quite unlike any other fungus; shaped like a tall egg, with the pointed end upwards, on a short, hollow stalk, and looking something like a sponge. It has a delicate and excellent flavour, and is perfectly wholesome.

The pretty little Woodruff is in flower; what scent is so delicate as that of its leaves? They are almost sweeter when dried, each little whorl by itself, with the stalk cut closely away above and below. It is a pleasant surprise to come upon these fragrant little stars between the leaves of a book. The whole plant revives memories of rambles in Bavarian woodlands, and of Maitrank, that best of the "cup" tribe of pleasant drinks, whose flavour is borrowed from its flowering tips.

In the first week in May oak-timber is being felled. The wood is handsomer, from showing the grain better, when it is felled in the winter, but it is delayed till now because of the value of the bark for tanning, and just now the fast-rising sap makes the bark strip easily. A heavy fall is taking place in the fringes of a large wood of old Scotch fir. Where the oaks grow there is a blue carpet of wild Hyacinth; the pathway is a slightly hollowed lane, so that the whole sheet of flower right and left is nearly on a level with the eye, and looks like solid pools of blue. The oaks not yet felled are putting forth their leaves of golden bronze. The song of the nightingale and the ring of the woodman's axe gain a rich musical quality from the great fir wood. Why a wood of Scotch fir has this wonderful property of a kind of musical reverberation I do not know; but so it is. Any sound that occurs within it is, on a lesser scale, like a sound in a cathedral. The tree itself when struck gives a musical note. Strike an oak or an elm on the trunk with a stick, and the

TRILLIUM IN THE WILD GARDEN.

sound is mute; strike a Scotch fir, and it is a note of music.

In the copse are some prosperous patches of the beautiful North American Wood-lily (*Trillium grandiflorum*). It likes a bed of deep leaf-soil on levels or cool slopes in woodland, where its large white flowers and whorls of handsome leaves look quite at home. Beyond it are widely spreading patches of Solomon's Seal and tufts of the Wood-rush (*Luzula sylvatica*), showing by their happy vigour how well they like their places, while the natural woodland carpet of moss and dead leaves puts the whole together. Higher in copse the path runs through stretches of the pretty little *Smilacina bifolia*, and the ground beyond this is a thick bed of Whortleberry, filling all the upper part of the copse under oak and birch and Scotch fir. The little flower-bells of the Whortleberry have already given place to the just-formed fruit, which will ripen in July, and be a fine feast for the blackbirds.

Other parts of the copse, where there was no Heath or Whortleberry, were planted thinly with the large Lily of the Valley. It has spread and increased and become broad sheets of leaf and bloom, from which thousands of flowers can be gathered without making gaps, or showing that any have been removed; when the bloom is over the leaves still stand in handsome masses till they are hidden by the fast-growing bracken. They do not hurt each other, as it seems that the Lily of the Valley, having the roots running just underground, while the fern-roots are much deeper, the two occupy their respective *strata* in perfect good fellowship. The neat little *Smilacina* is a near relation of the Lily of the Valley ; its leaves are of an even more vivid green, and its little modest spikes of white flower are charming. It loves the poor, sandy soil, and increases in it fast, but will have nothing to say to clay. A very delicate and beautiful North American fern (*Dicksonia punctilobulata*) proves a good colonist in the copse. It spreads rapidly by creeping roots, and looks much like our native *Thelipteris*, but is of a paler green colour. In the rock-garden the brightest patches of bloom are shown by the tufts of dwarf Wallflowers ; of these, *Cheiranthus alpinus* has a strong lemon colour that is of great brilliancy in the mass, and *C. Marshalli* is of a dark orange colour, equally powerful. The curiously-tinted *C. mutabilis*, as its name implies, changes from a light mahogany colour when just open, first to crimson and then to purple. In length of life *C. alpinus* and *C. Marshalli* are rather more than biennials, and yet too short-lived to be called true perennials ; cuttings of one year flower the next, and are handsome tufts the year after, but are scarcely worth keeping longer. *C. mutabilis* is longer lived, especially if the older growths are cut right away, when the tuft will generally spring into vigorous new life.

Orobus aurantiacus is a beautiful plant not enough grown, one of the handsomest of the Pea family, with flowers of a fine orange colour, and foliage of a healthy-looking golden-green. A striking and handsome plant in the upper part of the rockery is *Othonna cheirifolia;* its aspect is unusual and interesting, with its bunches of thick, blunt-edged leaves of blue-grey colouring, and large yellow daisy flowers. There is a pretty group of the large white Thrift, and near it a spreading carpet of blue Veronica and some of the splendid gentian-blue *Phacelia campanularia,* a valuable annual for filling any bare patches of rockery where its brilliant colouring will suit the neighbouring plants, or, best of all, in patches among dwarf ferns, where its vivid blue would be seen to great advantage.

Two wall-shrubs have been conspicuously beautiful during May; the Mexican Orange-flower (*Choisya ternata*) has been smothered in its white bloom, so closely resembling orange-blossom. With a slight winter protection of fir boughs it seems quite at home in our hot, dry, soil, grows fast, and is very easy to propagate by layers. When cut, it lasts for more than a week in water. *Piptanthus nepalensis* has also made a handsome show, with its abundant yellow, pea-shaped bloom and deep-green trefoil leaves. The dark-green stems have a slight bloom on a half-polished surface, and a pale ring at each joint gives them somewhat the look of bamboos.

Now is the time to look out for the big queen wasps and to destroy as many as possible. They seem to be specially fond of the flowers of two plants, the

large perennial Cornflower (*Centaurea montana*) and the
common Cotoneaster. I have often secured a dozen in
a few minutes on one or other of these plants, first
knocking them down with a battledore.

Now, in the third week of May, Rhododendrons
are in full bloom on the edge of the copse. The plantation was made about nine years ago, in one of the
regions where lawn and garden were to join the wood.
During the previous blooming season the best nurseries
were visited and careful observations made of colouring, habit, and time of blooming. The space they
were to fill demanded about seventy bushes, allowing
an average of eight feet from plant to plant—not
seventy different kinds, but, perhaps, ten of one kind,
and two or three fives, and some threes, and a few
single plants, always bearing in mind the ultimate
intention of pictorial aspect as a whole. In choosing
the plants and in arranging and disposing the groups
these ideas were kept in mind: to make pleasant ways
from lawn to copse; to group only in beautiful colour
harmonies; to choose varieties beautiful in themselves;
to plant thoroughly well, and to avoid overcrowding.
Plantations of these grand shrubs are generally spoilt
or ineffective, if not absolutely jarring, for want of
attention to these simple rules. The choice of kinds
is now so large, and the variety of colouring so exten-
sive, that nothing can be easier than to make beautiful
combinations, if intending planters will only take the
small amount of preliminary trouble that is needful.

RHODODENDRONS WHERE THE COPSE AND GARDEN MEET.

Some of the clumps are of brilliant scarlet-crimson, rose
and white, but out of the great choice of colours that
might be so named only those are chosen that make
just the colour-harmony that was intended. A large
group, quite detached from this one, and more in the
shade of the copse, is of the best of the lilacs, purples,
and whites. When some clumps of young hollies
have grown, those two groups will not be seen at the
same time, except from a distance. The purple and
white group is at present rather the handsomest, from
the free-growing habit of the fine old kind *Album elegans*,
which forms towering masses at the back. A detail
of pictorial effect that was aimed at, and that has
come out well, was devised in the expectation that
the purple groups would look richer in the shade, and
the crimson ones in the sun. This arrangement has
answered admirably. Before planting, the ground, of
the poorest quality possible, was deeply trenched, and
the Rhododendrons were planted in wide holes filled
with peat, and finished with a comfortable "mulch," or
surface - covering of farmyard manure. From this a
supply of grateful nutriment was gradually washed
in to the roots. This beneficial surface-dressing was
renewed every year for two years after planting, and
even longer in the case of the slower growing kinds.
No plant better repays care during its early years.
Broad grass paths leading from the lawn at several
points pass among the clumps, and are continued
through the upper parts of the copse, passing through

E

zones of different trees; first a good stretch of birch and holly, then of Spanish chestnut, next of oak, and finally of Scotch fir, with a sprinkling of birch and mountain ash, all with an undergrowth of heath and whortleberry and bracken. Thirty years ago it was all a wood of old Scotch fir. This was cut at its best marketable maturity, and the present young wood is made of what came up self-sown. This natural wild growth was thick enough to allow of vigorous cutting out, and the preponderance of firs in the upper part and of birch in the lower suggested that these were the kinds that should predominate in their respective places.

It may be useful to describe a little more in detail the plan I followed in grouping Rhododendrons, for I feel sure that any one with a feeling for harmonious colouring, having once seen or tried some such plan, will never again approve of the haphazard mixtures. There may be better varieties representing the colourings aimed at in the several groups, but those named are ones that I know, and they will serve as well as any others to show what is meant.

The colourings seem to group themselves into six classes of easy harmonies, which I venture to describe thus :—

1. Crimsons inclining to scarlet or blood-colour grouped with dark claret-colour and true pink.

In this group I have planted Nigrescens, dark claret-colour ; John Waterer and James Marshall Brook,

GRASS WALKS THROUGH THE COPSE.

both fine red-crimsons; Alexander Adie and Atrosan-guineum, good crimsons, inclining to blood-colour; Alarm, rosy-scarlet; and Bianchi, pure pink.

2. Light scarlet rose colours inclining to salmon, a most desirable range of colour, but of which the only ones I know well are Mrs. R. S. Holford, and a much older kind, Lady Eleanor Cathcart. These I put by themselves, only allowing rather near them the good pink Bianchi.

3. Rose colours inclining to amaranth.

4. Amaranths or magenta-crimsons.

5. Crimson or amaranth-purples.

6. Cool clear purples of the typical *ponticum* class, both dark and light, grouped with lilac-whites, such as *Album elegans* and *Album grandiflorum*. The beauti-ful partly-double *Everestianum* comes into this group, but nothing redder among purples. *Fastuosum flore-pleno* is also admitted, and *Luciferum* and *Reine Hor-tense*, both good lilac-whites. But the purples that are most effective are merely *ponticum* seedlings, chosen when in bloom in the nursery for their depth and rich-ness of cool purple colour.

My own space being limited, I chose three of the above groups only, leaving out, as of colouring less pleasing to my personal liking, groups 3, 4, and 5. The remaining ones gave me examples of colouring the most widely different, and at the same time the most agreeable to my individual taste. It would have been easier, if that had been the object, to have made groups of the three other classes of colouring, which comprise by far the largest number of the splendid varieties now grown.   There are a great many beautiful whites; of these, two that I most admire are Madame Carvalho and Sappho; the latter is an immense flower, with a conspicuous purple blotch.   There is also a grand old kind called Minnie, a very large-growing one, with fine white trusses; and a dwarf-growing white that comes early into bloom is Cunningham's White, also useful for forcing, as it is a small plant, and a free bloomer.

Nothing is more perplexing than to judge of the relative merits of colours in a Rhododendron nursery, where they are all mixed up.   I have twice been specially to look for varieties of a true pink colour, but the quantity of untrue pinks is so great that anything approaching a clear pink looks much better than it is. In this way I chose Kate Waterer and Sylph, both splendid varieties; but when I grew them with my true pink Bianchi they would not do, the colour having the suspicion of rank quality that I wished to keep out of that group.   This same Bianchi, with its mongrel-sounding name, I found was not grown in the larger nurseries.   I had it from Messrs. Maurice Young, of the Milford Nurseries, near Godalming.   I regretted to hear lately from some one to whom I recommended it that it could not be supplied.   It is to be hoped that so good a thing has not been lost.

A little way from the main Rhododendron clumps, and among bushy Andromedas, I have the splendid

RHODODENDRONS AT THE EDGE OF THE COPSE.

hybrid of *R. Aucklandi*, raised by Mr. A. Waterer
The trusses are astoundingly large, and the individual blooms large and delicately beautiful, like
small richly-modelled lilies of a tender, warm, white
colour. It is quite hardy south of London, and un-
questionably desirable. Its only fault is leggy growth;
one year's growth measures twenty-three inches, but
this only means that it should be planted among other
bushes.

The last days of May see hardy Azaleas in beauty
Any of them may be planted in company, for all their
colours harmonise. In this garden, where care is taken
to group plants well for colour, the whites are planted
at the lower and more shady end of the group; next
come the pale yellows and pale pinks, and these are
followed at a little distance by kinds whose flowers are
of orange, copper, flame, and scarlet-crimson colourings; this strong-coloured group again softening off
at the upper end by strong yellows, and dying away
into the woodland by bushes of the common yellow
Azalea pontica, and its variety with flowers of larger
size and deeper colour. The plantation is long in
shape, straggling over a space of about half an acre,
the largest and strongest-coloured group being in
an open clearing about midway in the length. The
ground between them is covered with a natural growth
of the wild Ling (*Calluna*) and Whortleberry, and the
small, white-flowered Bed-straw, with the fine-bladed
Sheep's-fescue grass, the kind most abundant in heath-

land. The surrounding ground is copse, of a wild,
forest-like character, of birch and small oak. A wood-
path of wild heath cut short winds through the planted
group, which also comprises some of the beautiful
white - flowered Californian *Azalea occidentalis*, and
bushes of some of the North American Vacciniums.

Azaleas should never be planted among or even
within sight of Rhododendrons. Though both enjoy
a moist peat soil, and have a near botanical relationship, they are incongruous in appearance, and impossible
to group together for colour. This must be understood
to apply to the two classes of plants of the hardy
kinds, as commonly grown in gardens. There are
tender kinds of the East Indian families that are quite
harmonious, but those now in question are the ordinary
varieties of so-called Ghent Azaleas, and the hardy
hybrid Rhododendrons. In the case of small gardens,
where there is only room for one bed or clump of peat
plants, it would be better to have a group of either
one or the other of these plants, rather than spoil the
effect by the inharmonious mixture of both.

I always think it desirable to group together
flowers that bloom at the same time. It is impossible,
and even undesirable, to have a garden in blossom all
over, and groups of flower-beauty are all the more enjoyable for being more or less isolated by stretches of
intervening greenery. As one lovely group for May I
recommend Moutan Pæony and *Clematis montana*, the
Clematis on a wall low enough to let its wreaths of

bloom show near the Pæony. The old Guelder Rose
or Snowball-tree is beautiful anywhere, but I think it
best of all on the cold side of a wall. Of course it is
perfectly hardy, and a bush of strong, sturdy growth,
and has no need of the wall either for support or for
shelter; but I am for clothing the garden walls with
all the prettiest things they can wear, and no shrub I
know makes a better show. Moreover, as there is
necessarily less wood in a flat wall tree than in a round
bush, and as the front shoots must be pruned close
back, it follows that much more strength is thrown into
the remaining wood, and the blooms are much larger.

I have a north wall eleven feet high, with a Guelder Rose on each side of a doorway, and a *Clematis
montana* that is trained on the top of the whole. The
two flower at the same time, their growths mingling
in friendly fashion, while their unlikeness of habit
makes the companionship all the more interesting.
The Guelder Rose is a stiff-wooded thing, the character
of its main stems being a kind of stark uprightness,
though the great white balls hang out with a certain
freedom from the newly-grown shoots. The Clematis
meets it with an exactly opposite way of growth,
swinging down its great swags of many-flowered garland masses into the head of its companion, with here
and there a single flowering streamer making a tiny
wreath on its own account.

On the southern sides of the same gateway are two
large bushes of the Mexican Orange-flower (*Choisya*

ternata), loaded with its orange-like bloom. Buttresses flank the doorway on this side, dying away into the general thickness of the wall above the arch by a kind of roofing of broad flat stones that lay back at an easy pitch. In mossy hollows at their joints and angles, some tufts of Thrift and of little Rock Pinks have found a home, and show as tenderly-coloured tufts of rather dull pink bloom. Above all is the same white Clematis, some of its abundant growth having been trained over the south side, so that this one plant plays a somewhat important part in two garden-scenes.

Through the gateway again, beyond the wall northward and partly within its shade, is a portion of ground devoted to Pæonies, in shape a long triangle, whose proportion in length is about thrice its breadth measured at the widest end. A low cross-wall, five feet high, divides it nearly in half near the Guelder Roses, and it is walled again on the other long side of the triangle by a rough structure of stone and earth, which, in compliment to its appearance, we call the Old Wall, of which I shall have something to say later. Thus the Pæonies are protected all round, for they like a sheltered place, and the Moutans do best with even a little passing shade at some time of the day. Moutan is the Chinese name for Tree Pæony. For an immense hardy flower of beautiful colouring what can equal the salmon-rose Moutan Reine Elizabeth ? Among the others that I have, those that give me most pleasure are Baronne d'Alès

SOUTH SIDE OF DOOR, WITH CLEMATIS MONTANA AND CHOISYA.

NORTH SIDE OF THE SAME DOOR, WITH CLEMATIS MONTANA
AND GUELDER-ROSE.

and Comtesse de Tuder, both pinks of a delightful quality, and a lovely white called Bijou de Chusan. The Tree Pæonies are also beautiful in leaf; the individual leaves are large and important, and so carried that they are well displayed. Their colour is peculiar, being bluish, but pervaded with a suspicion of pink or pinkish-bronze, sometimes of a metallic quality that faintly recalls some of the variously-coloured alloys of metal that the Japanese bronze-workers make and use with such consummate skill.

It is a matter of regret that varieties of the better kinds of Moutans are not generally grown on their own roots, and still more so that the stock in common use should not even be the type Tree Pæony, but one of the herbaceous kinds, so that we have plants of a hard-wooded shrub worked on a thing as soft as a Dahlia root. This is probably the reason why they are so difficult to establish, and so slow to grow, especially on light soils, even when their beds have been made deep and liberally enriched with what one judges to be the most gratifying comfort. Every now and then, just before blooming time, a plant goes off all at once, smitten with sudden death. At the time of making my collection I was unable to visit the French nurseries where these plants are so admirably grown, and whence most of the best kinds have come. I had to choose them by the catalogue description—always an unsatisfactory way to any one with a keen eye for colour, although in this matter the compilers of foreign catalogues are

certainly less vague than those of our own. Many of
the plants therefore had to be shifted into better
groups for colour after their first blooming, a matter
the more to be regretted as Pæonies dislike being
moved.

The other half of the triangular bit of Pæony ground
—the pointed end—is given to the kinds I like best
of the large June-flowered Pæonies, the garden varieties
of the Siberian *P. albiflora,* popularly known as Chinese
Pæonies. Though among these, as is the case with
all the kinds, there is a preponderance of pink or rose-
crimson colouring of a decidedly rank quality, yet the
number of varieties is so great, that among the minority
of really good colouring there are plenty to choose
from, including a good number of beautiful whites and
whites tinged with yellow. Of those I have, the kinds
I like best are—

Hypatia, pink.
Madame Benare, salmon-rose.
The Queen, pale salmon-rose.
Léonie, salmon-rose.
Virginie, warm white.
Solfaterre, pale yellow.
Edouard André, deep claret.
Madame Calot, flesh pink.
Madame Bréon.
Alba sulfurea.
Triomphans gandavensis.
Carnea elegans (Guerin).
Curiosa, pink and blush.

Prince Pierre Galitzin, blush.
Eugenie Verdier, pale pink.
Elegans superbissima, yellowish-
    white.
Virgo Maria, white.
Philomèle, blush.
Madame Dhour, rose.
Duchesse de Nemours, yellow-
    white.
Faust.
Belle Donaisienne.
Jeanne d'Arc.
Marie Lemonie.

Many of the lovely flowers in this class have a rather

strong, sweet smell, something like a mixture of the scents of Rose and Tulip.

Then there are the old garden Pæonies, the double varieties of *P. officinalis.* They are in three distinct colourings—full rich crimson, crimson-rose, and pale pink changing to dull white. These are the earliest to flower, and with them it is convenient, from the garden point of view, to class some of the desirable species.

Some years ago my friend Mr. Barr kindly gave me a set of the Pæony species as grown by him. I wished to have them, not for the sake of making a collection, but in order to see which were the ones I should like best to grow as garden flowers. In due time they grew into strong plants and flowered. A good many had to be condemned because of the raw magenta colour of the bloom, one or two only that had this defect being reprieved on account of their handsome foliage and habit. Prominent among these was *P. decora,* with bluish foliage handsomely displayed, the whole plant looking strong and neat and well-dressed. Others whose flower-colour I cannot commend, but that seemed worth growing on account of their rich masses of handsome foliage, are *P. triternata* and *P. Broteri.* Though small in size, the light red flower of *P. lobata* is of a beautiful colour. *P. tenuifolia,* in both single and double form, is an old garden favourite. *P. Wittmanniana,* with its yellow-green leave sand tender yellow flower, is a gem ; but it is rather rare, and probably uncertain, for mine, alas !

had no sooner grown into a fine clump than it suddenly died.

All Pæonies are strong feeders. Their beds should be deeply and richly prepared, and in later years they are grateful for liberal gifts of manure, both as surface dressings and waterings.

Friends often ask me vaguely about Pæonies, and when I say, " What kind of Pæonies ? " they have not the least idea.

Broadly, and for garden purposes, one may put them into three classes—

1. Tree Pæonies (*P. moutan*), shrubby, flowering in May.

2. Chinese Pæonies (*P. albiflora*), herbaceous, flowering in June.

3. Old garden Pæonies (*P. officinalis*), herbaceous, including some other herbaceous species.

I find it convenient to grow Pæony species and Caulescent (Lent) Hellebores together. They are in a wide border on the north side of the high wall and partly shaded by it. They are agreed in their liking for deeply-worked ground with an admixture of loam and lime, for shelter, and for rich feeding; and the Pæony clumps, set, as it were, in picture frames of the lower-growing Hellebores, are seen to all the more advantage.

Free Cluster-Rose as standard in a Cottage Garden.

CHAPTER VII

JUNE

The gladness of June—The time of Roses—Garden Roses—Reine Blanche—The old white Rose—Old garden Roses as standards —Climbing and rambling Roses—Scotch Briars—Hybrid Perpetuals a difficulty — Tea Roses — Pruning — Sweet Peas, autumn sown — Elder-trees — Virginian Cowslip — Dividing spring-blooming plants—Two best Mulleins—White French Willow—Bracken.

WHAT is one to say about June—the time of perfect young summer, the fulfilment of the promise of the earlier months, and with as yet no sign to remind one that its fresh young beauty will ever fade? For my own part I wander up into the wood and say, "June is here—June is here; thank God for lovely June!" The soft cooing of the wood-dove, the glad song of many birds, the flitting of butterflies, the hum of all the little winged people among the branches, the sweet earth-scents—all seem to say the same, with an endless reiteration, never wearying because so gladsome. It is the offering of the Hymn of Praise! The lizards run in and out of the heathy tufts in the hot sunshine, and as the long day darkens the night-jar trolls out his strange song, so welcome because it is the prelude to the perfect summer night; here and there a glow-worm shows its little lamp. June is here—June is here; thank God for lovely June!

And June is the time of Roses. I have great delight in the best of the old garden Roses; the Provence (Cabbage Rose), sweetest of all sweets, and the Moss Rose, its crested variety; the early Damask, and its red and white striped kind; the old, nearly single, Reine Blanche. I do not know the origin of this charming Rose, but by its appearance it should be related to the Damask. A good many years ago I came upon it in a cottage garden in Sussex, and thought I had found a white Damask. The white is a creamy white, the outsides of the outer petals are stained with red, first showing clearly in the bud. The scent is delicate and delightful, with a faint suspicion of Magnolia. A few years ago this pretty old Rose found its way to one of the meetings of the Royal Horticultural Society, where it gained much praise. It was there that I recognised my old friend, and learned its name.

I am fond of the old *Rosa alba*, both single and double, and its daughter, Maiden's Blush. How seldom one sees these Roses except in cottage gardens; but what good taste it shows on the cottager's part, for what Rose is so perfectly at home upon the modest little wayside porch?

I have also learnt from cottage gardens how pretty are some of the old Roses grown as standards. The

picture of my neighbour, Mrs. Edgeler, picking me a
bunch from her bush, shows how freely they flower,
and what fine standards they make. I have taken the
hint, and have now some big round-headed standards,
the heads a yard through, of the lovely Celeste and of
Madame Plantier, that are worth looking at, though
one of them is rather badly-shaped this year, for my
handsome Jack (donkey) ate one side of it when he
was waiting outside the studio door, while his cart-load
of logs for the ingle fire was being unloaded.

What a fine thing, among the cluster Roses, is the
old Dundee Rambler! I trained one to go up a rather
upright green Holly about twenty-five feet high, and
now it has rushed up and tumbles out at the top and
sides in masses of its pretty bloom.   It is just as good
grown as a "fountain," giving it a free space where it
can spread at will with no training or support whatever.   These two ways I think are much the best for
growing the free, rambling Roses. In the case of
the fountain, the branches arch over and display the
flowers to perfection; if you tie your Rose up to a
tall post or train it over an arch or *pergola*, the birds
flying overhead have the best of the show.   The
Garland Rose, another old sort, is just as suitable for
this kind of growth as Dundee Rambler, and the
individual flowers, of a tender blush-colour, changing
to white, are even more delicate and pretty.

The newer Crimson Rambler is a noble plant for
the same use, in sunlight gorgeous of bloom, and always

brilliant with its glossy bright-green foliage. Of the many good plants from Japan, this is the best that has reached us of late years. The Himalayan *Rosa Brunonii* is loaded with its clusters of milk-white bloom, that are so perfectly in harmony with its very long, almost blue leaves. But of all the free-growing Roses, the most remarkable for rampant growth is *R. Polyantha*. One of the bushes in this garden covers a space thirty-four feet across—more than a hundred feet round. It forms a great fountain-like mass, covered with myriads of its small white flowers, whose scent is carried a considerable distance. Directly the flower is over it throws up rods of young growth eighteen to twenty feet long; as they mature they arch over, and next year their many short lateral shoots will be smothered with bloom.

Two other Roses of free growth are also great favourites—Madame Alfred Carrière, with long-stalked loose white flowers, and Emilie Plantier. I have them on an east fence, where they yield a large quantity of bloom for cutting; indeed, they have been so useful in this way that I have planted several more, but this time for training down to an oak trellis, like the one that supports the row of Bouquet d'Or, in order to bring the flowers within easier reach.

Now we look for the bloom of the Burnet Rose (*Rosa spinosissima*), a lovely native plant, and its garden varieties, the Scotch Briars. The wild plant is widely distributed in England, though somewhat local. It

DOUBLE WHITE SCOTCH BRIAR.

grows on moors in Scotland, and on Beachy Head in
Sussex, and near Tenby in South Wales, favouring
wild places within smell of the sea. The rather dusky
foliage sets off the lemon-white of the wild, and the
clear white, pink, rose, and pale yellow of the double
garden kinds. The hips are large and handsome,
black and glossy, and the whole plant in late autumn
assumes a fine bronzy colouring between ashy black
and dusky red. Other small old garden Roses are
coming into bloom. One of the most desirable, and
very frequent in this district, is *Rosa lucida*, with red
stems, highly-polished leaves, and single, fragrant flowers
of pure rosy-pink colour. The leaves turn a brilliant
yellow in autumn, and after they have fallen the bushes
are still bright with the coloured stems and the large
clusters of bright-red hips. It is the St. Mark's Rose
of Venice, where it is usually in flower on St. Mark's
Day, April 25th. The double variety is the old *Rose
d'amour*, now rare in gardens; its half-expanded bud is
perhaps the most daintily beautiful thing that any Rose
can show.

After many years of fruitless effort I have to allow
that I am beaten in the attempt to grow the Grand
Roses in the Hybrid Perpetual class. They plainly
show their dislike to our dry hill, even when their
beds are as well enriched as I can contrive or afford
to make them. The rich loam that they love has to
come many miles from the Weald by hilly roads in
four-horse waggons, and the haulage is so costly that

F

when it arrives I feel like distributing it with a spoon
rather than with the spade. Moreover, even if a
bed is filled with the precious loam, unless constantly
watered the plants seem to feel and resent the two
hundred feet of dry sand and rock that is under them
before any moister stratum is reached.

But the Tea Roses are more accommodating, and do
fairly well, though, of course, not so well as in a stiffer
soil.   If I were planting again I should grow a still
larger proportion of the kinds I have now found to do
best.   Far beyond all others is Madame Lambard, good
alike early and late, and beautiful at all times.   In this
garden it yields quite three times as much bloom as
any other; nothing else can approach it either for
beauty or bounty.   Viscountess Folkestone, not properly
a Tea, but classed among Hybrid Noisettes, is also free
and beautiful and long-enduring ; and Papa Gontier, so
like a deeper-coloured Lambard, is another favourite.
Bouquet d'Or is here the strongest of the Dijon Teas.
I grow it in several positions, but most conveniently on
a strong bit of oak post and rail trellis, keeping the
long growths tied down, and every two years cutting the
oldest wood right out.   It is well to remember that
the tying or pegging down of Roses always makes them
bloom better : every joint from end to end wants to
make a good Rose ; if the shoots are more upright, the
blooming strength goes more to the top.

The pruning of Tea Roses is quite different from
the pruning required for the Hybrid Perpetuals.   In

PART OF A BUSH OF ROSA POLYANTHA.

GARLAND-ROSE, SHOWING NATURAL WAY OF GROWTH.

these the last year's growth is cut back in March to within two to five eyes from where it leaves the main branch, according to the strength of the kind. This must not be done with the Teas. With these the oldest wood is cut right out from the base, and the blooming shoots left full length. But it is well, towards the end of July or beginning of August, to cut back the ends of soft summer shoots in order to give them a chance of ripening what is left. When an old Tea looks worn out, if cut right down in March or April it will often throw out vigorous young growth, and quite renew its life.

Within the first days of June we can generally pick some Sweet Peas from the rows sown in the second week of September. They are very much stronger than those sown in spring. By November they are four inches high, and seem to gain strength and sturdiness during the winter; for as soon as spring comes they shoot up with great vigour, and we know that the spray used to support them must be two feet higher than for those that are spring-sown. The flower-stalks are a foot long, and many have four flowers on a stalk. They are sown in shallow trenches; in spring they are earthed up very slightly, but still with a little trench at the base of the plants. A few doses of liquid manure are a great help when they are getting towards blooming strength.

I am very fond of the Elder-tree. It is a sociable sort of thing; it seems to like to grow near human habitations. In my own mind it is certainly the tree most closely associated with the pretty old cottage and farm architecture of my part of the country; no bush or tree, not even the apple, seems to group so well or so closely with farm buildings. When I built a long thatched shed for the many needs of the garden, in the region of pits and frames, compost, rubbish and burn-heap, I planted Elders close to the end of the building and on one side of the yard. They look just right, and are, moreover, every year loaded with their useful fruit. This is ripe quite early in September, and is made into Elder wine, to be drunk hot in winter, a comfort by no means to be despised. My trees now give enough for my own wants, and there are generally a few acceptable bushels to spare for my cottage neighbours.

About the middle of the month the Virginian Cowslip (*Mertensia virginica*) begins to turn yellow before dying down. Now is the time to look out for the seeds. A few ripen on the plant, but most of them fall while green, and then ripen in a few days while lying on the ground. I shake the seeds carefully out, and leave them lying round the parent-plant; a week later, when they will be ripe, they are lightly scratched into the ground. Some young plants of last year's growth I mark with a bit of stick, in case of wanting some later to plant elsewhere, or to send away; the plant dies away completely, leaving no trace above ground, so that if not marked it would be difficult to find what is wanted.

LILAC MARIE LEGRAYE. (*See page* 23.)

FLOWERING ELDER AND PATH FROM GARDEN TO COPSE.

This is also the time for pulling to pieces and replenishing that good spring plant, the large variety of *Myosotis dissitiflora;* I always make sure of divisions, as seed does not come true. *Primula rosea* should also be divided now, and planted to grow on in a cool place, such as the foot of a north or east wall, or be put at once in its place in some cool, rather moist spot in the rock-garden. Two-year-old plants come up with thick clumps of matted root that is now useless. I cut off the whole mass of old root about an inch below the crown, when it can easily be divided into nice little bits for replanting. Many other spring - flowering plants may with advantage be divided now, such as Aubrietia, Arabis, Auricula, Tiarella, and Saxifrage.

The young Primrose plants, sown in March, have been planted out in their special garden, and are looking well after some genial rain.

The great branching Mullein, *Verbascum olympicum,* is just going out of bloom, after making a brilliant display for a fortnight. It is followed by the other of the most useful tall, yellow-flowered kinds, *V. phlomoides.* Both are seen at their best either quite early in the morning, or in the evening, or in half-shade, as, like all their kind, they do not expand their bloom in bright sunshine. Both are excellent plants on poor soils. *V. olympicum,* though classed as a biennial, does not come to flowering strength till it is three or four years old; but meanwhile the foliage is so handsome that even if there were no flower it would be a worthy garden plant. It does well in any waste spaces of poor soil, where, by having plants of all ages, there will be some to flower every year. The Mullein moth is sure to find them out, and it behoves the careful gardener to look for and destroy the caterpillars, or he may some day find, instead of his stately Mulleins, tall stems only clothed with unsightly grey rags. The caterpillars are easily caught when quite small or when rather large ; but midway in their growth, when three-quarters of an inch long, they are wary, and at the approach of the avenging gardener they will give a sudden wriggling jump, and roll down into the lower depths of the large foliage, where they are difficult to find. But by going round the plants twice a day for about a week they can all be discovered.

The white variety of the French Willow (*Epilobium angustifolium*) is a pretty plant in the edges of the copse, good both in sun and shade, and flourishing in any poor soil. In better ground it grows too rank, running quickly at the root and invading all its neighbours, so that it should be planted with great caution ; but when grown on poor ground it flowers at from two feet to four feet high, and its whole aspect is improved by the proportional amount of flower becoming much larger.

Towards the end of June the bracken that covers the greater part of the ground of the copse is in full beauty. No other manner of undergrowth gives to woodland in so great a degree the true forest-like character. This most ancient plant speaks of the old, untouched land of which large stretches still remain in the south of England—land too poor to have been worth cultivating, and that has therefore for centuries endured human contempt. In the early part of the present century, William Cobbett, in his delightful book, " Rural Rides," speaking of the heathy headlands and vast hollow of Hindhead, in Surrey, calls it " certainly the most villainous spot God ever made." This gives expression to his view, as farmer and political economist, of such places as were incapable of cultivation, and of the general feeling of the time about lonely roads in waste places, as the fields for the lawless labours of smuggler and highwayman. Now such tracts of natural wild beauty, clothed with stretches of Heath and Fern and Whortleberry, with beds of Sphagnum Moss, and little natural wild gardens of curious and beautiful sub-aquatic plants in the marshy hollows and undrained wastes, are treasured as such places deserve to be, especially when they still remain within fifty miles of a vast city. The height to which the bracken grows is a sure guide to the depth of soil. On the poorest, thinnest ground it only reaches a foot or two ; but in hollow places where leaf-mould accumulates and surface soil has washed in and made a better depth, it grows from six feet to eight feet high, and when straggling up through bushes to get to the light a frond will sometimes measure as much as twelve feet. The old country people who have always lived

on the same poor land say, " Where the farn grows tall
anything will grow "; but that only means that there
the ground is somewhat better and capable of cultivation, as its presence is a sure indication of a sandy soil.
The timber-merchants are shy of buying oak trees
felled from among it, the timber of trees grown on the
wealden clay being so much better.

CHAPTER VIII

JULY

Scarcity of flowers—Delphiniums—Yuccas—Cottager's way of protecting tender plants—Alströmerias—Carnations—Gypsophila —*Lilium giganteum*—Cutting fern-pegs.

AFTER the wealth of bloom of June, there appear to be but few flowers in the garden; there seems to be a time of comparative emptiness between the earlier flowers and those of autumn. It is true that in the early days of July we have Delphiniums, the grandest blues of the flower year. They are in two main groups in the flower border, one of them nearly all of the palest kind—not a solid clump, but with a thicker nucleus, thinning away for several yards right and left. Only white and pale-yellow flowers are grouped with this, and pale, fresh-looking foliage of maize and Funkia. The other group is at some distance, at the extreme western end. This is of the full and deeper blues, following a clump of Yuccas, and grouped about with things of important silvery foliage, such at Globe Artichoke and Silver Thistle (*Eryngium*). I have found it satisfactory to grow Delphiniums from seed, choosing the fine strong " Cantab " as the seed-parent,

because the flowers were of a medium colour—scarcely
so light as the name would imply—and because of its
vigorous habit and well-shaped spike.   It produced
flowers of all shades of blue, and from these were
derived nearly all I have in the border.   I found them
better for the purpose in many cases than the named
kinds of which I had a fair collection.

The seedlings were well grown for two years in
nursery lines, worthless ones being taken out as soon
as they showed their character.   There is one common
defect that I cannot endure—an interrupted spike,
when the flowers, having filled a good bit of the spike,
leave off, leaving a space of bare stem, and then go on
again.   If this habit proves to be persistent after the
two years' trial, the plant is condemned.   For my
liking the spike must be well filled, but not over-
crowded.   Many of the show kinds are too full for
beauty; the shape of the individual flower is lost.
Some of the double ones are handsome, but in these
the flower takes another shape, becoming more rosette-
like, and thereby loses its original character.   Some
are of mixed colouring, a shade of lilac-pink sliding
through pale blue.   It is very beautiful in some cases,
the respective tints remaining as clear as in an opal,
but in many it only muddles the flower and makes it
ineffective.

Delphiniums are greedy feeders, and pay for rich
cultivation and for liberal manurial mulches and
waterings.   In a hot summer, if not well cared for,

they get stunted and are miserable objects, the flower distorted and cramped into a clumsy-looking, elongated mop-head.

Though weak in growth the old *Delphinium Bella-donna* has so lovely a quality of colour that it is quite indispensable ; the feeble stem should be carefully and unobtrusively staked for the better display of its incomparable blue.

Some of the Yuccas will bloom before the end of the month. I have them in bold patches the whole fifteen-feet depth of the border at the extreme ends, and on each side of the pathway, where, passing from the lawn to the Pæony ground, it cuts across the border to go through the arched gateway.   The kinds of Yucca are *gloriosa, recurva, flaccida,* and *filamentosa.* They are good to look at at all times of the year because of their grand strong foliage, and are the glory of the garden when in flower.   One of the *gloriosa* threw up a stout flower-spike in January.   I had thought of protecting and roofing the spike, in the hope of carrying it safely through till spring, but meanwhile there came a damp day and a frosty night, and when I saw it again it was spoilt.   The *Yucca filamentosa* that I have I was told by a trusty botanist was the true plant, but rather tender, the one commonly called by that name being something else.   I found it in a cottage garden, where I learnt a useful lesson in protecting plants, namely, the use of thickly-cut peaty sods.   The goodwife had noticed that the peaty ground of the adjoining common, covered with heath and gorse and mossy grass, resisted frost much better than the garden or meadow, and it had been her practice for many years to get some thick dry sods with the heath left on and to pack them close round to protect tender plants. In this way she had preserved her Fuchsias of greenhouse kinds, and Calceolarias, and the Yucca in question.

The most brilliant mass of flower in early July is given by the beds of *Alströmeria aurantiaca*; of this we have three distinct varieties, all desirable. There is a four feet wide bed, some forty feet long, of the kind most common in gardens, and at a distance from it a group grown from selected seed of a paler colour; seedlings of this remain true to colour, or, as gardeners say, the variety is " fixed." The third sort is from a good old garden in Ireland, larger in every way than the type, with petals of great width, and extremely rich in colour. *Alströmeria chilense* is an equally good plant, and beds of it are beautiful in their varied colourings, all beautifully harmonious, and ranging through nearly the same tints as hardy Azaleas. These are the best of the Alströmerias for ordinary garden culture; they do well in warm, sheltered places in the poorest soil, but the soil must be deep, for the bunches of tender, fleshy roots go far down. The roots are extremely brittle, and must be carefully handled. Alströmerias are easily raised from seed, but when the seedlings are planted out the crowns should be

quite four inches under the surface, and have a thick
bed of leaves or some other mild mulching material
over them in winter to protect them from frost, for
they are Chilian plants, and demand and deserve a
little surface comfort to carry them safely through the
average English winter.

Sea-holly (*Eryngium*) is another family of July-
flowering plants that does well on poor, sandy soils
that have been deeply stirred. Of these the more
generally useful is *E. Olivieranum*, the *E. amethystinum*
of nurserymen, but so named in error, the true plant
being rare and scarcely known in gardens. The whole
plant has an admirable structure of a dry and nervous
quality, with a metallic colouring and dull lustre that
are in strong contrast to softer types of vegetation.
The black-coated roots go down straight and deep, and
enable it to withstand almost any drought. Equalling
it in beauty is *E. giganteum*, the Silver Thistle, of the
same metallic texture, but whitish and almost silvery.
This is a biennial, and should be sown every year.
A more lowly plant, but hardly less beautiful, is the
wild Sea-holly of our coasts (*E. maritimum*), with leaves
almost blue, and a handsome tuft of flower nearly
matching them in colour. It occurs on wind-blown
sandhills, but is worth a place in any garden. It comes
up rather late, but endures, apparently unchanged,
except for the bloom, throughout the late summer
and autumn.

But the flower of this month that has the firmest

hold of the gardener's heart is the Carnation—the
Clove Gilliflower of our ancestors. Why the good old
name " Gilliflower " has gone out of use it is impossible
to say, for certainly the popularity of the flower has
never waned. Indeed, in the seventeenth century it
seems that it was the best-loved flower of all in England ; for John Parkinson, perhaps our earliest writer
on garden plants, devotes to it a whole chapter in his
" Paradisus Terrestris," a distinction shared by no other
flower. He describes no less than fifty kinds, a few
of which are still to be recognised, though some are
lost. For instance, what has become of the " *great gray
Hulo*," which he describes as a plant of the largest and
strongest habit ? The " gray " in this must refer to
the colour of the leaf, as he says the flower is red ; but
there is also a variety called the " *blew Hulo*," with
flowers of a " purplish murrey " colouring, answering to
the slate colour that we know as of not unfrequent
occurrence. The branch of the family that we still
cultivate as " Painted Lady " is named by him " Dainty
Lady," the present name being no doubt an accidental
and regrettable corruption. But though some of the
older sorts may be lost, we have such a wealth of good
known kinds that this need hardly be a matter of
regret. The old red Clove always holds its own for
hardiness, beauty, and perfume ; its newer and dwarfer
variety, Paul Engleheart, is quite indispensable, while
the beautiful salmon-coloured Raby is perhaps the
most useful of all, with its hardy constitution and great

quantity of bloom. But it is difficult to grow Carnations on our very poor soil; even when it is carefully
prepared they still feel its starving and drying influence, and show their distaste by unusual shortness of
life.

Gypsophila paniculata is one of the most useful
plants of this time of year; its delicate masses of
bloom are like clouds of flowery mist settled down
upon the flower borders. Shooting up behind and
among it is a tall, salmon-coloured Gladiolus, a telling
contrast both in form and manner of inflorescence.
Nothing in the garden has been more satisfactory
and useful than a hedge of the white everlasting Pea.
The thick, black roots that go down straight and deep
have been undisturbed for some years, and the plants
yield a harvest of strong white bloom for cutting that
always seems inexhaustible. They are staked with stiff,
branching spray, thrust into the ground diagonally,
and not reaching up too high. This supports the
heavy mass of growth without encumbering the upper
blooming part.

Hydrangeas are well in flower at the foot of a warm
wall, and in the same position are spreading masses of
the beautiful *Clematis Davidiana*, a herbaceous kind,
with large, somewhat vine-like leaves, and flowers of a
pale-blue colour of a delicate and uncommon quality.

The blooming of the *Lilium giganteum* is one of the
great flower events of the year. It is planted in rather
large straggling groups just within the fringe of the

copse.   In March the bulbs, which are only just underground, thrust their sharply-pointed bottle-green tips
out of the earth.   These soon expand into heart-shaped
leaves, looking much like Arum foliage of the largest
size, and of a bright-green colour and glistening surface.   The groups are so placed that they never see
the morning sun.   They require a slight sheltering
of fir-bough, or anything suitable, till the third week
of May, to protect the young leaves from the late
frosts.   In June the flower-stem shoots up straight
and tall, like a vigorous young green-stemmed tree.
If the bulb is strong and the conditions suitable, it
will attain a height of over eleven feet, but among the
flowering bulbs of a group there are sure to be some
of various heights from differently sized bulbs; those
whose stature is about ten feet are perhaps the handsomest.   The upper part of the stem bears the gracefully drooping great white Lily flowers, each bloom
some ten inches long, greenish when in bud, but changing to white when fully developed.   Inside each petal
is a purplish-red stripe.   In the evening the scent seems
to pour out of the great white trumpets, and is almost
overpowering, but gains a delicate quality by passing
through the air, and at fifty yards away is like a faint
waft of incense.   In the evening light, when the sun
is down, the great heads of white flower have a mysterious and impressive effect when seen at some distance
through the wood, and by moonlight have a strangely
weird dignity.   The flowers only last a few days, but

THE GIANT LILY.

when they are over the beauty of the plant is by no means gone, for the handsome leaves remain in perfection till the autumn, while the growing seed-pods, rising into an erect position, become large and rather handsome objects. The rapidity and vigour of the four months' growth from bulb to giant flowering plant is very remarkable. The stem is a hollow, fleshy tube, three inches in diameter at the base, and the large radiating roots are like those of a tree. The original bulb is, of course, gone, but when the plants that have flowered are taken up at the end of November, offsets are found clustered round the root; these are carefully detached and replanted. The great growth of these Lilies could not be expected to come to perfection in our very poor, shallow soil, for doubtless in their mountain home in the Eastern Himalayas they grow in deep beds of cool vegetable earth. Here, therefore, their beds are deeply excavated, and filled to within a foot of the top with any of the vegetable rubbish of which only too much accumulates in the late autumn. Holes twelve feet across and three feet deep are convenient graves for frozen Dahlia-tops and half-hardy Annuals; a quantity of such material chopped up and tramped down close forms a cool subsoil that will comfort the Lily bulbs for many a year. The upper foot of soil is of good compost, and when the young bulbs are planted, the whole is covered with some inches of dead leaves that join in with the natural woodland carpet.

In the end of July we have some of the hottest of

G

the summer days, only beginning to cool between six
and seven in the evening. One or two evenings I go
to the upper part of the wood to cut some fern-pegs
for pegging Carnation layers, armed with fag-hook and
knife and rubber, and a low rush-bottomed stool to sit
on. The rubber is the stone for sharpening the knife—
a long stone of coarse sandstone grit, such as is used
for scythes. Whenever I am at work with a knife
there is sure to be a rubber not far off, for a blunt
knife I cannot endure, so there is a stone in each
department of the garden sheds, and a whole series in
the workshop, and one or two to spare to take on outside jobs. The Bracken has to be cut with a light
hand, as the side-shoots that will make the hook of the
peg are easily broken just at the important joint. The
fronds are of all sizes, from two to eight feet long; but
the best for pegs are the moderate-sized, that have not
been weakened by growing too close together. Where
they are crowded the main stalk is thick, but the side
ones are thin and weak; whereas, where they get light
and air the side branches are carried on stouter ribs,
and make stronger and better-balanced pegs. The cut
fern is lightly laid in a long ridge with the ends all
one way, and the operator sits at the stalk end of the
ridge, a nice cool shady place having been chosen.
Four cuts with the knife make a peg, and each frond
makes three pegs in about fifteen seconds. With the
fronds laid straight and handy it goes almost rhyth-
mically, then each group of three pegs is thrown into

the basket, where they clash on to the others with a
hard ringing sound.   In about four days the pegs dry
to a surprising hardness; they are better than wooden
ones, and easier and quicker to make.

People who are not used to handling Bracken
should be careful how they cut a frond with a knife;
they are almost sure to get a nasty little cut on the
second joint of the first finger of the right hand—not
from the knife, but from the cut edge of the fern.
The stalk has a silicious coating, that leaves a sharp
edge like a thin flake of glass when cut diagonally
with a sharp knife; they should also beware how they
pick or pull off a mature frond, for even if the part of
the stalk laid hold of is bruised and twisted, some of
the glassy structure holds together, and is likely to
wound the hand.

CHAPTER IX

AUGUST

Leycesteria formosa is a soft-wooded shrub, whose beauty,
without being showy, is full of charm and refinement.
I remember delighting in it in the shrub-wilderness of
the old home, where I first learnt to know and love
many a good bush and tree long before I knew their
names. There were towering Rhododendrons (all *ponticum*) and Ailantus and Hickory and Magnolias, and
then Spiræa and Snowball tree and tall yellow Azalea,
and Buttercup bush and shrubby Andromedas, and in
some of the clumps tall Cypresses and the pretty cut-
leaved Beech, and in the edges of others some of the
good old garden Roses, double Cinnamon and *R. lucida*,
and Damask and Provence, Moss-rose and Sweetbriar,
besides tall-grown Lilacs and Syringa. It was all
rather overgrown, and perhaps all the prettier, and
some of the wide grassy ways were quite shady in
summer. And I look back across the years and think

THE GREAT ASPHODEL.

CISTUS FLORENTINUS.

what a fine lesson-book it was to a rather solitary
child; and when I came to plant my own shrub
clump I thought I would put rather near together
some of the old favourites, so here again we come
back to Leycesteria, put rather in a place of honour,
and near it Buttercup bush and Andromeda and Magnolias and old garden Roses.

I had no space for a shrub wilderness, but have
made a large clump for just the things I like best,
whether new friends or old. It is a long, low bank,
five or six paces wide, highest in the middle, where
the rather taller things are planted. These are mostly
Junipers and Magnolias; of the Magnolias, the kinds
are *Soulangeana, conspicua, purpurea,* and *stellata.* One
end of the clump is all of peat earth; here are Andromedas, Skimmeas, and on the cooler side the broad-
leaved Gale, whose crushed leaves have almost the
sweetness of Myrtle. One long side of the clump
faces south-west, the better to suit the things that
love the sun. At the farther end is a thrifty bush of
Styrax japonica, which flowers well in hot summers,
but another bush under a south wall flowers better.
It must be a lovely shrub in the south of Europe and
perhaps in Cornwall; here the year's growth is always
cut at the tip, but it flowers well on the older wood,
and its hanging clusters of white bloom are lovely.
At its foot, on the sunny side, are low bushy plants of
Cistus florentinus. I am told that this specific name
is not right; but the plant so commonly goes by it

that it serves the purpose of popular identification. Then comes *Magnolia stellata*, now a perfectly-shaped bush five feet through, a sheet of sweet-scented bloom in April. Much too near it are two bushes of *Cistus ladaniferus*. They were put there as little plants to grow on for a year in the shelter and comfort of the warm bank, but were overlooked at the time they ought to have been shifted, and are now nearly five feet high, and are crowding the Magnolia. I cannot bear to take them away to waste, and they are much too large to transplant, so I am driving in some short stakes diagonally and tying them down by degrees, spreading out their branches between neighbouring plants. It is an upright-growing Cistus that would soon cover a tallish wall-space, but this time it must be content to grow horizontally, and I shall watch to see whether it will flower more freely, as so many things do when trained down.

Next comes a patch of the handsome *Bambusa Ragamowski*, dwarf, but with strikingly-broad leaves of a bright yellow-green colour. It seems to be a slow grower, or more probably it is slow to grow at first; Bamboos have a good deal to do underground. It was planted six years ago, a nice little plant in a pot, and now is eighteen inches high and two feet across. Just beyond it is the Mastic bush (*Cary-opteris mastacanthus*), a neat, grey-leaved small shrub, crowded in September with lavender - blue flowers, arranged in spikes something like a Veronica; the whole bush is aromatic, smelling strongly like highly-refined turpentine. Then comes *Xanthoceras sorbifolia*, a handsome bush from China, of rather recent introduction, with saw-edged pinnate leaves and white flowers earlier in the summer, but now forming its bunches of fruit that might easily be mistaken for walnuts with their green shucks on. Here a wide bushy growth of *Phlomis fruticosa* lays out to the sun, covered in early summer with its stiff whorls of hooded yellow flowers—one of the best of plants for a sunny bank in full sun in a poor soil. A little farther along, and near the path, comes the neat little *Deutzia parviflora* and another little shrub of fairy-like delicacy, *Philadelphus microphyllus*. Behind them is *Stephanandra flexuosa*, beautiful in foliage, and two good St. John's worts, *Hypericum aureum* and *H. Moserianum*, and again in front a Cistus of low, spreading growth, *C. halimifolius*, or something near it. One or two favourite kinds of Tree Pæonies, comfortably sheltered by Lavender bushes, fill up the other end of the clump next to the Andromedas. In all spare spaces on the sunny side of the shrub-clump is a carpeting of *Megasea ligulata*, a plant that looks well all the year round, and gives a quantity of precious flower for cutting in March and April.

I was nearly forgetting *Pavia macrostachya*, now well established among the choice shrubs. It is like a bush Horse-chestnut, but more refined, the white spikes standing well up above the handsome leaves.

On the cooler side of the clump is a longish planting of dwarf Andromeda, precious not only for its
beauty of form and flower, but from the fine winter
colouring of the leaves, and those two useful Spiræas,
S. Thunbergi, with its countless little starry flowers,
and the double *prunifolia*, the neat leaves of whose
long sprays turn nearly scarlet in autumn. Then
there comes a rather long stretch of *Artemisia stel-
leriana*, a white-leaved plant much like *Cineraria
maritima*, answering just the same purpose, but perfectly hardy. It is so much like the silvery *Cineraria*
that it is difficult to remember that it prefers a cool
and even partly-shaded place.

Beyond the long ridge that forms the shrub-
clump is another, parallel to it and only separated
from it by a path, also in the form of a long low
bank. On the crown of this is the double row of
cob-nuts that forms one side of the nut-alley. It
leaves a low sunny bank that I have given to various
Briar Roses and one or two other low, bushy kinds.
Here is the wild Burnet Rose, with its yellow-white
single flowers and large black hips, and its garden
varieties, the Scotch Briars, double white, flesh-coloured,
pink, rose, and yellow, and the hybrid briar, Stanwell
Perpetual. Here also is the fine hybrid of *Rosa rugosa*,
Madame George Bruant, and the lovely double *Rosa
lucida*, and one or two kinds of small bush Roses from
out-of-the-way gardens, and two wild Roses that have
for me a special interest, as I collected them from

LAVENDER HEDGE AND STEPS TO THE LOFT.

HOLLYHOCK, PINK BEAUTY.

their rocky home in the island of Capri. One is a Sweetbriar, in all ways like the native one, except that the flowers are nearly white, and the hips are larger. Last year the bush was distinctly more showy than any other of its kind, on account of the size and unusual quantity of the fruit. The other is a form of *Rosa sempervirens*, with rather large white flowers faintly tinged with yellow.

Hollyhocks have been fine, in spite of the disease, which may be partly checked by very liberal treatment. By far the most beautiful is one of a pure pink colour, with a wide outer frill. It came first from a cottage garden, and has always since been treasured. I call it Pink Beauty. The wide outer petal (a heresy to the florist) makes the flower infinitely more beautiful than the all-over full-double form that alone is esteemed on the show-table. I shall hope in time to come upon the same shape of flower in white, sulphur, rose-colour, and deep blood-crimson, the colours most worth having in Hollyhocks.

Lavender has been unusually fine; to reap its fragrant harvest is one of the many joys of the flower year. If it is to be kept and dried, it should be cut when as yet only a few of the purple blooms are out on the spike; if left too late, the flower shakes off the stalk too readily.

Some plantations of *Lilium Harrisi* and *Lilium auratum* have turned out well. Some of the *Harrisi* were grouped among tufts of the bright-foliaged *Funkia grandiflora* on the cool side of a Yew hedge. Just at the foot of the hedge is *Tropæolum speciosum*, which runs up into it and flowers in graceful wreaths some feet above the ground. The masses of pure white lily and cool green foliage below are fine against the dark, solid greenery of the Yew, and the brilliant flowers above are like little jewels of flame. The Bermuda Lilies (*Harrisi*) are intergrouped with *L. speciosum*, which will follow them when their bloom is over. The *L. auratum* were planted among groups of Rhododendrons; some of them are between tall Rhododendrons, and have large clumps of Lady Fern (*Filix fœmina*) in front, but those that look best are between and among Bamboos (*B. Metake*); the heavy heads of flower borne on tall stems bend gracefully through the Bamboos, which just give them enough support.

Here and there in the copse, among the thick masses of green Bracken, is a frond or two turning yellow. This always happens in the first or second week of August, though it is no indication of the approaching yellowing of the whole. But it is taken as a signal that the Fern is in full maturity, and a certain quantity is now cut to dry for protection and other winter uses. Dry Bracken lightly shaken over frames is a better protection than mats, and is almost as easily moved on and off.

The Ling is now in full flower, and is more beautiful in the landscape than any of the garden Heaths; the

SOLOMON'S SEAL IN SPRING, IN THE UPPER PART OF THE FERN-WALK.

THE FERN-WALK IN AUGUST.

relation of colouring, of greyish foliage and low-toned pink bloom with the dusky spaces of purplish-grey shadow, are a precious lesson to the colour-student.

The fern-walk is at its best. It passes from the garden upwards to near the middle of the copse. The path, a wood-path of moss and grass and short-cut heath, is a little lower than the general level of the wood. The mossy bank, some nine feet wide, and originally cleared for the purpose, is planted with large groups of hardy Ferns, with a preponderance (due to preference) of Dilated Shield Fern and Lady Fern. Once or twice in the length of the bank are hollows, sinking at their lowest part to below the path-level, for *Osmunda* and *Blechnum*. When rain is heavy enough to run down the path it finds its way into these hollow places.

Among the groups of Fern are a few plants of true wood-character—*Linnœa*, *Trientalis*, *Goodyera*, and *Trillium*. At the back of the bank, and stretching away among the trees and underwood, are wide-spreading groups of Solomon's-seal and Wood-rush, joining in with the wild growth of Bracken and Bramble.

Most of the Alpines and dwarf-growing plants, whose home is the rock-garden, bloom in May or June, but a few flower in early autumn. Of these one of the brightest is *Ruta patavina*, a dwarf plant with lemon-coloured flowers and a very neat habit of growth. It soon makes itself at home in a sunny bank in poor soil. *Pterocephalus parnassi* is a dwarf Scabious, with small, grey foliage keeping close to the ground, and rather large flowers of a low-toned pink. The white Thyme is a capital plant, perfectly prostrate, and with leaves of a bright yellow-green, that with the white bloom give the plant a particularly fresh appearance. It looks at its best when trailing about little flat spaces between the neater of the hardy Ferns, and hanging over little rocky ledges. Somewhat farther back is the handsome dwarf *Platycodon Mariesi*, and behind it the taller Platycodons, among full-flowered bushes of *Olearia Haasti*.

By the middle of August the garden assumes a character distinctly autumnal. Much of its beauty now depends on the many non-hardy plants, such as Gladiolus, Canna, and Dahlia, on Tritomas of doubtful hardiness, and on half-hardy annuals — Zinnia, Helichrysum, Sunflower, and French and African Marigold. Fine as are the newer forms of hybrid Gladiolus, the older strain of gandavensis hybrids are still the best as border flowers. In the large flower border, tall, well-shaped spikes of a good pink one look well shooting up through and between a wide-spreading patch of glaucous foliage of the smaller Yuccas, *Tritoma glaucescens*, *Iris pallida*, and *Funkia Sieboldi*, while scarlet and salmon-coloured kinds are among groups of Pæonies that flowered in June, whose leaves are now taking a fine reddish colouring. Between these and the edge of the border is a straggling group some yards in length of the dark-foliaged

Heuchera Richardsoni, that will hold its satin-surfaced leaves till the end of the year. Farther back in the border is a group of the scarlet-flowered Dahlia Fire King, and behind these, Dahlias Lady Ardilaun and Cochineal, of deeper scarlet colouring. The Dahlias are planted between groups of Oriental Poppy, that flower in May and then die away till late in autumn. Right and left of the scarlet group are Tritomas, intergrouped with Dahlias of moderate height, and with orange and flame-coloured flowers. This leads to some masses of flowers of strong yellow colouring; the old perennial Sunflower, in its tall single form, and the best variety of the old double one of moderate height, the useful *H. lætiflorus* and the tall Miss Mellish, the giant form of *Harpalium rigidum*. *Rudbekia Newmanni* reflects the same strong colour in the front part of the border, and all spaces are filled with orange Zinnias and African Marigolds and yellow Helichrysum. As we pass along the border the colour changes to paler yellow by means of a pale perennial Sunflower and the sulphur-coloured annual kind, with Paris Daisies, *Œnothera Lamarkiana* and *Verbascum phlomoides*. The two last were cut down to about four feet after their earliest bloom was over, and are now again full of profusely-flowered lateral growths. At the farther end of the border we come again to glaucous foliage and pale-pink flower of Gladiolus and Japan Anemone. It is important in such a border of rather large size, that can be seen from a good space

of lawn, to keep the flowers in rather large masses of
colour.   No one who has ever done it, or seen it done,
will go back to the old haphazard sprinkle of colouring without any thought of arrangement, such as is
usually seen in a mixed border.   There is a wall of
sandstone backing the border, also planted in relation
to the colour-massing in the front space.   This gives
a quiet background of handsome foliage, with always
in the flower season some show of colour in one part
or another of its length.   Just now the most conspicuous of its clothing shrubs or of the somewhat tall
growing flowers at its foot are a fine variety of *Bignonia
radicans*, a hardy Fuchsia, the Claret Vine covering a
good space, with its red-bronze leaves and clusters of
blue-black grapes, the fine hybrid Crinums and *Clerodendron fœtidum.*

Tea Roses have been unusually lavish of autumn
bloom, and some of the garden climbing Roses, hybrids
of China and Noisette, have been of great beauty, both
growing and as room decoration.   Many of them flower
in bunches at the end of the shoots; whole branches,
cut nearly three feet long, make charming arrangements in tall glasses or high vases of Oriental china.
Perhaps their great autumnal vigour is a reaction
from the check they received in the earlier part of the
year, when the bloom was almost a failure from the
long drought and the accompanying attacks of blight
and mildew.   The great hips of the Japanese *Rosa
rugosa* are in perfection; they have every ornamental

quality—size, form, colour, texture, and a delicate waxlike bloom; their pulp is thick and luscious, and makes an excellent jam.

The quantity of fungous growth this year is quite remarkable. The late heavy rain coming rather suddenly on the well-warmed earth has no doubt brought about their unusual size and abundance; in some woodland places one can hardly walk without stepping upon them. Many spots in the copse are brilliant with large groups of the scarlet-capped Fly Agaric (*Amanita muscaria*). It comes out of the ground looking like a dark scarlet ball, generally flecked with raised whitish spots; it quickly rises on its white stalk, the ball changing to a brilliant flat disc, six or seven inches across, and lasting several days in beauty. But the most frequent fungus is the big brown *Boletus*, in size varying from a small bun to a dinner-plate. Some kinds are edible, but I have never been inclined to try them, being deterred by their coarse look and uninviting coat of slimy varnish. And why eat doubtful *Boletus* when one can have the delicious Chantarelle (*Cantharellus cibarius*), also now at its best? In colour and smell it is like a ripe apricot, perfectly wholesome, and, when rightly cooked, most delicate in flavour and texture. It should be looked for in cool hollows in oak woods; when once found and its good qualities appreciated, it will never again be neglected.

CHAPTER X

SEPTEMBER

In the second week of September we sow Sweet Peas in shallow trenches. The flowers from these are larger and stronger and come in six weeks earlier than from those sown in the spring; they come too at a time when they are especially valuable for cutting. Many other hardy Annuals are best sown now. Some indeed, such as the lovely *Collinsia verna* and the large white Iberis, only do well if autumn-sown. Among others, some of the most desirable are Nemophila, Platystemon, Love-in-a-Mist, Larkspurs, Pot Marigold, Virginian Stock, and the delightful Venus's Navel-wort (*Omphalodes lini-folia*). I always think this daintily beautiful plant is undeservedly neglected, for how seldom one sees it. It is full of the most charming refinement, with its milk-white bloom and grey-blue leaf and neat habit of growth. Any one who has never before tried Annuals autumn-sown would be astonished at their vigour. A single plant of Nemophila will often cover a square yard with its beautiful blue bloom; and then, what a gain it is to have these pretty things in full strength in spring and early summer, instead of waiting to have them in a much poorer state later in the year, when other flowers are in plenty.

Hardy Poppies should be sown even earlier; August is the best time.

Dahlias are now at their full growth. To make a choice for one's own garden, one must see the whole plant growing. As with many another kind of flower, nothing is more misleading than the evidence of the show-table, for many that there look the best, and are indeed lovely in form and colour as individual blooms, come from plants that are of no garden value. For however charming in humanity is the virtue modesty, and however becoming is the unobtrusive bearing that gives evidence of its possession, it is quite misplaced in a Dahlia. Here it becomes a vice, for the Dahlia's first duty in life is to flaunt and to swagger and to carry gorgeous blooms well above its leaves, and on no account to hang its head. Some of the delicately-coloured kinds lately raised not only hang their heads, but also hide them away among masses of their coarse foliage, and are doubly frauds, looking everything that is desirable in the show, and proving worthless in the garden. It is true that there are ways of cutting out superfluous green stuff and thereby encouraging the blooms to show up, but at a busy

H

season, when rank leafage grows fast, one does not want
to be every other day tinkering at the Dahlias.

Careful and strong staking they must always have,
not forgetting one central stake to secure the main
growth at first. It is best to drive this into the hole
made for the plant before placing the root, to avoid
the danger of sending the point of the stake through
the tender tubers. Its height out of the ground
should be about eighteen inches less than the expected
stature of the plant. As the Dahlia grows, there
should be at least three outer stakes at such distance
from the middle one as may suit the bulk and habit
of the plant; and it is a good plan to have wooden
hoops to tie to these, so as to form a girdle round the
whole plant, and for tying out the outer branches.
The hoop should be only loosely fastened—best with
roomy loops of osier, so that it may be easily shifted
up with the growth of the plant. We make the hoops
in the winter of long straight rent rods of Spanish
Chestnut, bending them while green round a tub, and
tying them with tarred twine or osier bands. They
last several years. All this care in staking the Dahlias
is labour well bestowed, for when autumn storms come
the wind has such a power of wrenching and twisting,
that unless the plant, now grown into a heavy mass
of succulent vegetation, is braced by firm fixing at the
sides, it is in danger of being broken off short just
above the ground, where its stem has become almost
woody, and therefore brittle.

Now is the moment to get to work on the rock-garden; there is no time of year so precious for this work as September. Small things planted now, while the ground is still warm, grow at the root at once, and get both anchor-hold and feeding-hold of the ground before frost comes. Those that are planted later do not take hold, and every frost heaves them up, sometimes right out of the ground. Meanwhile those that have got a firm root-hold are growing steadily all the winter, underground if not above; and when the first spring warmth comes they can draw upon the reserve of strength they have been hoarding up, and make good growth at once.

Except in the case of a rockery only a year old, there is sure to be some part that wants to be worked afresh, and I find it convenient to do about a third of the space every year. Many of the indispensable Alpines and rock-plants of lowly growth increase at a great rate, some spreading over much more than their due space, the very reason of this quick-spreading habit being that they are travelling to fresh pasture; many of them prove it clearly by dying away in the middle of the patch, and only showing vigorous vitality at the edges.

Such plants as *Silene alpestris*, *Hutchinsia alpina*, *Pterocephalus*, the dwarf alpine kinds of *Achillea* and *Artemisia*, *Veronica* and *Linaria*, and the mossy Saxifrages, in my soil want transplanting every two years, and the silvery Saxifrages every three years. As in

much else, one must watch what happens in one's
own garden. We practical gardeners have no absolute
knowledge of the constitution of the plant, still less
of the chemistry of the soil, but by the constant
exercise of watchful care and helpful sympathy we
acquire a certain degree of instinctive knowledge, which
is as valuable in its way, and probably more applicable to individual local conditions, than the tabulated
formulas of more orthodox science.

One of the best and simplest ways of growing rock-
plants is in a loose wall. In many gardens an abrupt
change of level makes a retaining wall necessary, and
when I see this built in the usual way as a solid
structure of brick and mortar—unless there be any
special need of the solid wall—I always regret that it
is not built as a home for rock-plants. An exposure
to north or east and the cool backing of a mass of
earth is just what most Alpines delight in. A dry
wall, which means a wall without mortar, may be anything between a wall and a very steep rock-work, and
may be built of brick or of any kind of local stone. I
have built and planted a good many hundred yards of
dry walling with my own hands, both at home and in
other gardens, and can speak with some confidence both
of the pleasure and interest of the actual making and
planting, and of the satisfactory results that follow.

The best example I have to show in my own
garden is the so-called "Old Wall," before mentioned.
It is the bounding and protecting fence of the Pæony

The "Old Wall."

Jack. (*See page* 79.)

ground on its northern side, and consists of a double
dry wall with earth between. An old hedge bank that
was to come away was not far off, within easy wheel-
ing distance. So the wall was built up on each side,
and as it grew, the earth from the hedge was barrowed
in to fill up. A dry wall needs very little foundation;
two thin courses underground are quite enough. The
point of most structural importance is to keep the
earth solidly trodden and rammed behind the stones
of each course and throughout its bulk, and every two
or three courses to lay some stones that are extra long
front and back, to tie the wall well into the bank. A
local sandstone is the walling material. In the pit it
occurs in separate layers, with a few feet of hard sand
between each. The lowest layer, sometimes thirty to
forty feet down, is the best and thickest, but that is
good building stone, and for dry walling we only want
" tops " or " seconds," the later and younger formations
of stone in the quarry. The very roughness and
almost rotten state of much of this stone makes it
all the more acceptable as nourishment and root-hold
to the tiny plants that are to grow in its chinks, and
that in a few months will change much of the rough
rock-surface to green growth of delicate vegetation.
Moreover, much of the soft sandy stone hardens by
exposure to weather; and even if a stone or two
crumbles right away in a few years' time, the rest will
hold firmly, and the space left will make a little cave
where some small fern will live happily.

The wall is planted as it is built with hardy Ferns—
Blechnum, Polypody, Hartstongue, *Adiantum*, *Ceterach*,
Asplenium, and *Ruta muraria*. The last three like
lime, so a barrow of old mortar-rubbish is at hand,
and the joint where they are to be planted has a layer
of their favourite soil. Each course is laid fairly level
as to its front top edge, stones of about the same
thickness going in course by course. The earth backing is then carefully rammed into the spaces at the
uneven backs of the stones, and a thin layer of earth
over the whole course, where the mortar would have
been in a built wall, gives both a " bed " for the next
row of stones and soil for the plants that are to grow
in the joints.

The face of the wall slopes backward on both sides,
so that its whole thickness of five feet at the bottom
draws in to four feet at the top. All the stones are
laid at a right angle to the plane of the inclination—
that is to say, each stone tips a little down at the back,
and its front edge, instead of being upright, faces a
little upward. It follows that every drop of gentle
rain that falls on either side of the wall is carried into
the joints, following the backward and downward pitch
of the stones, and then into the earth behind them.

The mass of earth in the middle of the wall gives
abundant root-room for bushes, and is planted with
bush Roses of three kinds, of which the largest mass
is of *Rosa lucida*. Then there is a good stretch of
Berberis ; then Scotch Briars, and in one or two

important places Junipers; then more Berberis, and
Ribes, and the common Barberry, and neat bushes of
Olearia Haastii.

The wall was built seven years ago, and is now
completely clothed. It gives me a garden on the top
and a garden on each side, and though its own actual
height is only 4½ feet, yet the bushes on the top make
it a sheltering hedge from seven to ten feet high.
One small length of three or four yards of the top
has been kept free of larger bushes, and is planted
on its northern edge with a very neat and pretty dwarf
kind of Lavender, while on the sunny side is a thriving
patch of the hardy Cactus (*Opuntia Raffinesquiana*).
Just here, in the narrow border at the foot of the wall,
is a group of the beautiful *Crinum Powelli*, while a
white Jasmine clothes the face of the wall right and
left, and rambles into the Barberry bushes just beyond.
It so happened that these things had been planted
close together because the conditions of the place were
likely to favour them, and not, as is my usual practice,
with any intentional idea of harmonious grouping. I
did not even remember that they all flower in July,
and at nearly the same time; and one day seeing them
all in bloom together, I was delighted to see the success
of the chance arrangement, and how pretty it all was,
for I should never have thought of grouping together
pink and lavender, yellow and white.

The northern face of the wall, beginning at its
eastern end, is planted thus: For a length of ten or

twelve paces there are Ferns, Polypody and Hartstongue, and a few *Adiantum nigrum*, with here and
there a Welsh Poppy. There is a clump of the wild
Stitchwort that came by itself, and is so pretty that I
leave it. At the foot of the wall are the same, but
more of the Hartstongue; and here it grows best, for
not only is the place cooler, but I gave it some loamy
soil, which it loves. Farther along the Hartstongue
gives place to the wild Iris (*I. fœtidissima*), a good long
stretch of it. Nothing, to my mind, looks better than
these two plants at the base of a wall on the cool side.
In the upper part of the wall are various Ferns, and
that interesting plant, Wall Pennywort (*Cotyledon um-
bilicus*). It is a native plant, but not found in this
neighbourhood; I brought it from Cornwall, where it
is so plentiful in the chinks of the granite stone-fences.
It sows itself and grows afresh year after year, though I
always fear to lose it in one of our dry summers. Next
comes the common London Pride, which I think quite
the most beautiful of the Saxifrages of this section. If
it was a rare thing, what a fuss we should make about
it! The place is a little dry for it, but all the same,
it makes a handsome spreading tuft hanging over the
face of the wall. When its pink cloud of bloom is at
its best, I always think it the prettiest thing in the
garden. Then there is the Yellow Everlasting (*Gna-
phalium orientale*), a fine plant for the upper edge of
the wall, and even better on the sunny side, and the
white form of *Campanula cœspitosa*, with its crowd of

ERINUS ALPINUS, CLOTHING STEPS IN ROCK-WALL.

delicate little white bells rising in June, from the
neatest foliage of tender but lively green. Then follow
deep-hanging curtains of Yellow Alyssum and of hybrid
rock Pinks. The older plants of Alyssum are nearly
worn out, but there are plenty of promising young seedlings in the lower joints.

Throughout the wall there are patches of Polypody
Fern, one of the best of cool wall-plants, its creeping
root-stock always feeling its way along the joints, and
steadily furnishing the wall with more and more of its
neat fronds; it is all the more valuable for being at its
best in early winter, when so few ferns are to be seen.
Every year, in some bare places, I sow a little seed of
Erinus alpinus, always trying for places where it will
follow some other kind of plant, such as a place where
rock Pink or Alyssum has been. All plants are the better
for this sort of change. In the seven years that the
wall has stood, the stones have become weathered, and
the greater part of the north side, wherever the stone
work shows it, is hoary with mosses, and looks as if it
might have been standing for a hundred years.

The sunny side is nearly clear of moss, and I have
planted very few things in its face, because the narrow
border at its foot is so precious for shrubs and plants
that like a warm, sheltered place. Here are several
Choisyas and Sweet Verbenas, also *Escallonia*, *Stuartia*,
and *Styrax*, and a long straggling group of some very
fine Pentstemons. In one space that was fairly clear
I planted a bit of Hyssop, an old sweet herb whose

scent I delight in ; it grows into a thick bush-like plant
full of purple flower in the late summer, when it attracts
quantities of bumble-bees.   It is a capital wall-plant,
and has sown its own seed, till there is a large patch
on the top and some in its face, and a broadly-spread-
ing group in the border below.   It is one of the plants
that was used in the old Tudor gardens for edgings ;
the growth is close and woody at the base, and is easily
clipped into shape.

The fierce gales and heavy rains of the last days
of September wrought sad havoc among the flowers.
Dahlias were virtually wrecked.   Though each plant
had been tied to three stakes, their masses of heavy
growth could not resist the wrenching and twisting
action of the wind, and except in a few cases where
they were well sheltered, their heads lay on the ground,
the stems broken down at the last tie.   If anything
about a garden could be disheartening, it would be its
aspect after such a storm of wind.   Wall shrubs, only
lately made safe, as we thought, have great gaps torn
out of them, though tied with tarred string to strong
iron staples, staples and all being wrenched out.   Everything looks battered, and whipped, and ashamed ;
branches of trees and shrubs lie about far from their
sources of origin ; green leaves and little twigs are
washed up into thick drifts ; apples and quinces, that
should have hung till mid-October, lie bruised and
muddy under the trees.   Newly-planted roses and
hollies have a funnel-shaped hole worked in the ground

at their base, showing the power of the wind to twist
their heads, and giving warning of a corresponding
disturbance of the tender roots. There is nothing to
be done but to look round carefully and search out all
disasters and repair them as well as may be, and to
sweep up the wreckage and rubbish, and try to forget
the rough weather, and enjoy the calm beauty of the
better days that follow, and hope that it may be long
before such another angry storm is sent. And indeed
a few quiet days of sunshine and mild temperature
work wonders. In a week one would hardly know that
the garden had been so cruelly torn about. Fresh
flowers take the place of bruised ones, and wholesome
young growths prove the enduring vitality of vegetable
life. Still we cannot help feeling, towards the end of
September, that the flower year is nearly at an end,
though the end is a gorgeous one, with its strong
yellow masses of the later perennial Sunflowers and
Marigolds, Goldenrod, and a few belated Gladioli; the
brilliant foliage of Virginian Creepers, the leaf-painting
of *Vitis Coignettii*, and the strong crimson of the Claret
Vine.

The Water-elder (*Viburnum opulus*) now makes a
brave show in the edge of the copse. It is without
doubt the most beautiful berry-bearing shrub of mid-
September. The fruit hangs in ample clusters from
the point of every branch and of every lateral twig, in
colour like the brightest of red currants, but with a
translucent lustre that gives each separate berry a

much brighter look ; the whole bush shows fine warm colouring, the leaves having turned to a rich red. Perhaps it is because it is a native that this grand shrub or small tree is generally neglected in gardens, and is almost unknown in nurserymen's catalogues. It is the parent of the well-known Guelder-Rose, which is merely its double-flowered form. But the double flower leaves no berry, its familiar white ball being formed of the sterile part of the flower only, and the foliage of the garden kind does not assume so bright an autumn colouring.

The nights are growing chilly, with even a little frost, and the work for the coming season of dividing and transplanting hardy plants has already begun. Plans are being made for any improvements or alterations that involve ground work. Already we have been at work on some broad grass rides through the copse that were roughly levelled and laid with grass last winter. The turf has been raised and hollows filled in, grass seed sown in bare patches, and the whole beaten and rolled to a good surface, and the job put out of hand in good time before the leaves begin to fall.

CHAPTER XI

OCTOBER

THE early days of October bring with them the best
bloom of the Michaelmas Daisies, the many beautiful
garden kinds of the perennial Asters. They have, as
they well deserve to have, a garden to themselves.
Passing along the wide path in front of the big flower
border, and through the pergola that forms its continuation, with eye and brain full of rich, warm colouring of flower and leaf, it is a delightful surprise to pass
through the pergola's last right-hand opening, and to
come suddenly upon the Michaelmas Daisy garden in
full beauty. Its clean, fresh, pure colouring, of pale
and dark lilac, strong purple, and pure white, among
masses of pale-green foliage, forms a contrast almost
startling after the warm colouring of nearly everything
else; and the sight of a region where the flowers are

fresh and newly opened, and in glad spring-like pro-
fusion, when all else is on the verge of death and
decay, gives an impression of satisfying refreshment
that is hardly to be equalled throughout the year.
Their special garden is a wide border on each side of a
path, its length bounded on one side by a tall hedge
of filberts, and on the other side by clumps of yew,
holly, and other shrubs. It is so well sheltered that
the strongest wind has its destructive power broken,
and only reaches it as a refreshing tree-filtered breeze.
The Michaelmas Daisies are replanted every year as
soon as their bloom is over, the ground having been
newly dug and manured. The old roots, which will
have increased about fourfold, are pulled or chopped
to pieces, nice bits with about five crowns being chosen
for replanting; these are put in groups of three to five
together. Tall-growing kinds like *Novi Belgi*, Robert
Parker, are kept rather towards the back, while those
of delicate and graceful habit, such as *Cordifolius elegans*
and its good variety Diana are allowed to come forward. The fine dwarf *Aster amellus* is used in rather
large quantity, coming quite to the front in some
places, and running in and out between the clumps of
other kinds. Good-sized groups of *Pyrethrum uliginosum* are given a place among the Asters, for though
of quite another family, they are Daisies, and bloom
at Michaelmas, and are admirable companions to the
main occupants of the borders. The only other plants
admitted are white Dahlias, the two differently striped

BORDERS OF MICHAELMAS DAISIES.

varieties of *Eulalia japonica*, the fresh green foliage of Indian Corn, and the brilliant light-green leafage of *Funkia grandiflora*. Great attention is paid to staking the Asters. Nothing is more deplorable than to see a neglected, overgrown plant, at the last moment, when already half blown down, tied up in a tight bunch to one stake. When we are cutting underwood in the copse in the winter, special branching spray is looked out for our Michaelmas Daisies and cut about four feet or five feet long, with one main stem and from two to five branches. Towards the end of June and beginning of July these are thrust firmly into the ground among the plants, and the young growths are tied out so as to show to the best advantage. Good kinds of Michaelmas Daisies are now so numerous that in selecting those for the special garden it is well to avoid both the ones that bloom earliest and also the very latest, so that for about three weeks the borders may show a well-filled mass of bloom.

The bracken in the copse stands dry and dead, but when leaves are fluttering down and the chilly days of mid-October are upon us, its warm, rusty colouring is certainly cheering; the green of the freshly grown mossy carpet below looks vividly bright by contrast. Some bushes of Spindle-tree (*Euonymus europæus*) are loaded with their rosy seed-pods; some are already burst, and show the orange-scarlet seeds—an audacity of colouring that looks all the brighter for the even, lustreless green of the leaves and of the green-barked twigs and stems.

The hardy Azaleas are now blazing masses of crimson, almost scarlet leaf; the old *A. pontica*, with its large foliage, is as bright as any. With them are grouped some of the North American Vacciniums and Andromedas, with leaves almost as bright. The ground between the groups of shrubs is knee-deep in heath. The rusty-coloured withered bloom of the wild heath on its purplish-grey masses and the surrounding banks of dead fern make a groundwork and background of excellent colour-harmony.

How seldom does one see Quinces planted for ornament, and yet there is hardly any small tree that better deserves such treatment. Some Quinces planted about eight years ago are now perfect pictures, their lissome branches borne down with the load of great, deep-yellow fruit, and their leaves turning to a colour almost as rich and glowing. The old English rather round-fruited kind with the smooth skin is the best both for flavour and beauty—a mature tree without leaves in winter has a remarkably graceful, arching, almost weeping growth. The other kind is of a rather more rigid form, and though its woolly-coated, pear-shaped fruits are larger and strikingly handsome, the whole tree has a coarser look, and just lacks the attractive grace of the other. They will do fairly well almost anywhere, though they prefer a rich, loamy soil and a cool, damp, or even swampy place. The Medlar is another of the small fruiting trees that is more neglected than it should be, as it well deserves a place

among ornamental shrubs. Here it is a precious thing
in the region where garden melts into copse. The
fruit-laden twigs are just now very attractive, and its
handsome leaves can never be passed without admiration. Close to the Medlars is a happy intergrowth
of the wild Guelder-Rose, still bearing its brilliant
clusters, a strong-growing and far-clambering garden
form of *Rosa arvensis*, full of red hips, Sweetbriar, and
Holly—a happy tangle of red-fruited bushes, all looking
as if they were trying to prove, in friendly emulation,
which can make the bravest show of red-berried wild-
flung wreath, or bending spray, or stately spire; while
at their foot the bright colour is repeated by the bending, berried heads of the wild Iris, opening like fantastic
dragons' mouths, and pouring out the red bead-like
seeds upon the ground; and, as if to make the picture
still more complete, the leaves of the wild Strawberry
that cover the ground with a close carpet have also
turned to a crimson, and here and there to an almost
scarlet colour.

During the year I make careful notes of any
trees or shrubs that will be wanted, either to come
from the nursery or to be transplanted within my
own ground, so as to plant them as early as possible.
Of the two extremes it is better to plant too early
than too late. I would rather plant deciduous trees
before the leaves are off than wait till after Christmas,
but of all planting times the best is from the middle
of October till the end of November, and the same

I

time is the best for all hardy plants of large or
moderate size.

I have no patience with slovenly planting. I like
to have the ground prepared some months in advance,
and when the proper time comes, to do the actual planting as well as possible. The hole in the already prepared ground is taken out so that the tree shall stand
exactly right for depth, though in this dry soil it is
well to make the hole an inch or two deeper, in order
to leave the tree standing in the centre of a shallow
depression, to allow of a good watering now and then
during the following summer. The hole must be
made wide enough to give easy space for the most
outward-reaching of the roots; they must be spread
out on all sides, carefully combing them out with the
fingers, so that they all lay out to the best advantage.
Any roots that have been bruised, or have broken or
jagged ends, are cut off with a sharp knife on the home-
ward side of the injury. Most gardeners when they
plant, after the first spadeful or two has been thrown
over the root, shake the bush with an up and down
joggling movement. This is useful in the case of plants
with a good lot of bushy root, such as Berberis, helping
to get the grains of earth well in among the root; but
in tree planting, where the roots are laid out flat, it is of
course useless. In our light soil, the closer and firmer
the earth is made round the newly-planted tree the
better, and strong staking is most important, in order to
save the newly-placed root from disturbance by dragging.

Some trees and shrubs one can only get from nurseries in pots. Such is usually the case with Ilex, Escallonia, and Cydonia. Such plants are sure to have the roots badly matted and twisted. The main root curls painfully round and round inside the imprisoning pot, but if it is a clever root it works its way out through the hole in the bottom, and even makes quite nice roots in the bed of ashes it has stood on. In this case, as these are probably its best roots, we do not attempt to pull it back through the hole, but break the pot to release it without hurt. If it is possible to straighten the pot-curled root, it is best to do so; in any case, the small fibrous ones can be laid out. Often the potful of roots is so hard and tight that it cannot be disentangled by the hand; then the only way is to soften it by gentle bumping on the bench, and then to disengage the roots by little careful digs all round with a blunt-pointed stick. If this is not done, and the plant is put in in its pot-bound state, it never gets on; it would have been just as well to throw it away at once.

Nine years ago a hedge of Lawson's Cypress was planted on one side of the kitchen garden. Three years later, when the trees had made some growth, I noticed in the case of three or four that they were quite bare of branches on one side all the way up for a width of about one-sixth of the circumference, leaving a smooth, straight, upright strip. Suspecting the cause, I had them up, and found in every case that the

root just below the bare strip had been doubled under
the stem, and had therefore been unable to do its share
of the work. Nothing could have pointed out more
clearly the defect in the planting.

There are cases where ground cannot be prepared
as one would wish, and where one has to get over the
difficulty the best way one can. Such a case occurred
when I had to plant some Yews and Savins right under
a large Birch-tree. The Birch is one of several large
ones that nearly surround the lawn. This one stands
just within the end of a large shrub-clump, near the
place of meeting of some paths with the grass and with
some planting; here some further planting was wanted
of dark-leaved evergreens. There is no tree more
ground-robbing than a Birch, and under the tree in
question the ground was dust-dry, extremely hard, and
nothing but the poorest sand. Looking at the foot of
a large tree one can always see which way the main
roots go, and the only way to get down any depth is
to go between these and not many feet away from the
trunk. Farther away the roots spread out and would
receive more injury. So the ground was got up the
best way we could, and the Yews and Savins planted.
Now, after some six years, they are healthy and dark-
coloured, and have made good growth. But in such a
place one cannot expect the original preparation of the
ground, such as it was, to go for much. The year after
planting they had some strong, lasting manure just
pricked in over the roots—stuff from the shoeing-forge,

full of hoof-parings. Hoof-parings are rich in ammonia, and decay slowly. Every other year they have either a repetition of this or some cooling cow manure. The big Birch no doubt gets some of it, though its hungriest roots are farther afield, but the rich colour of the shrubs shows that they are well nourished.

As soon as may be in November the big hardy flower-border has to be thoroughly looked over. The first thing is to take away all " soft stuff." This includes all dead annuals and biennials and any tender things that have been put in for the summer, also Paris Daisies, Zinnias, French and African Marigolds, Helichrysums, Mulleins, and a few Geraniums. Then Dahlias are cut down. The waste stuff is laid in big heaps on the edge of the lawn just across the footpath, to be loaded into the donkey-cart and shot into some large holes that have been dug up in the wood, whose story will be told later.

The Dahlias are now dug up from the border, and others collected from different parts of the garden. The labels are tied on to the short stumps that remain, and the roots are laid for a time on the floor of a shed. If the weather has been rainy just before taking them up, it is well to lay them upside down, so that any wet there may be about the bases of the large hollow stalks may drain out. They are left for perhaps a fortnight without shaking out the earth that holds between the tubers, so that they may be fairly dry before they are put away for the winter in a cellar.

Then we go back to the flower border and dig out all the plants that have to be divided every year. It will also be the turn for some others that only want division every two or three or more years, as the case may be. First, out come all the perennial Sunflowers. These divide themselves into two classes; those whose roots make close clumpy masses, and those that throw out long stolons ending in a blunt snout, which is the growing crown for next year. To the first division belong the old double Sunflower (*Helianthus multiflorus*), of which I only keep the well-shaped variety Soleil d'Or, and the much taller large-flowered single kind, and a tall pale-yellow flowered one with a dark stem, whose name I do not know. It is not one of the kinds thought much of, and as usually grown has not much effect; but I plant it at the back and pull it down over other plants that have gone out of flower, so that instead of having only a few flowers at the top of a rather bare stem eight feet high, it is a spreading cloud of pale yellow bloom; the training down, as in the case of so many other plants, inducing it to throw up a short flowering stalk from the axil of every leaf along the stem. The kinds with the running roots are *Helianthus rigidus*, and its giant variety Miss Mellish, *H. decapetalus* and *H. lætiflorus*. I do not know how it may be in other gardens, but in mine these must be replanted every year.

Phloxes must also be taken up. They are always difficult here, unless the season is unusually rainy;

in dry summers, even with mulching and watering, I cannot keep them from drying up. The outside pieces are cut off and the woody middle thrown away. It is surprising what a tiny bit of Phlox will make a strong flowering plant in one season. The kinds I like best are the pure whites and the salmon-reds; but two others that I find very pretty and useful are Eugénie, a good mauve, and Le Soleil, a strong pink, of a colour as near a really good pink as in any Phlox I know. Both of these have a neat and rather short habit of growth. I do not have many Michaelmas Daisies in the flower border, only some early ones that flower within September; of these there are the white-flowered *A. paniculatus, Shortii, acris,* and *amellus.* These of course come up, and any patches of Gladiolus are collected, to be dried for a time and then stored.

The next thing is to look through the border for the plants that require occasional renewal. In the front I find that a longish patch of *Heuchera Richardsoni* has about half the plants overgrown. These must come up, and are cut to pieces. It is not a nice plant to divide; it has strong middle crowns, and though there are many side ones, they are attached to the main ones too high up to have roots of their own; but I boldly slice down the main stocky stem with straight downward cuts, so as to give a piece of the thick stock to each side bit. I have done this both in winter and spring, and find the spring rather the best, if not followed by drought. Groups of *Anemone japonica* and of *Polygonum compactum* are spreading beyond bounds and must be reduced. Neither of these need be entirely taken up. Without going into further detail, it may be of use to note how often I find it advisable to lift and divide some of the more prominent hardy plants.

Every year I divide Michaelmas Daisies, Golden-rod, *Helianthus, Phlox, Chrysanthemum maximum, Helenium pumilum, Pyrethrum uliginosum, Anthemis tinctoria, Monarda, Lychnis, Primula*, except *P. denticulata, rosea*, and *auricula*, which stand two years.

Every two years, White Pinks, Cranesbills, *Spiræa, Aconitum, Gaillardia, Coreopsis, Chrysanthemum indicum, Galega, Doronicum, Nepeta, Geum aureum, Œnothera Youngi*, and *Œ. riparia*.

Every three years, *Tritoma, Megasea, Centranthus, Vinca, Iris, Narcissus*.

A plasterer's hammer is a tool that is very handy for dividing plants. It has a hammer on one side of the head, and a cutting blade like a small chopper on the other. With this and a cold chisel and a strong knife one can divide any roots in comfort. I never divide things by brutally chopping them across with a spade. Plants that have soft fleshy tubers like Dahlias and Pæonies want the cold chisel; it can be cleverly inserted among the crowns so that injury to the tubers is avoided, and it is equally useful in the case of some plants whose points of attachment are almost as hard as wire, like *Orobus vernus*, or as tough

as a door-mat, like *Iris gramineas*. The Michaelmas
Daisies of the *Novæ Angliæ* section make root tufts too
close and hard to be cut with a knife, and here the
chopper of the plasterer's hammer comes in. Where
the crowns are closely crowded, as in this Aster, I find
it best to chop at the bottom of the tuft, among the
roots; when the chopper has cut about two-thirds
through, the tuft can be separated with the hands,
dividing naturally between the crowns, whereas if
chopped from the top many crowns would have been
spoilt.

Tritomas want dividing with care; it always looks
as if one could pull every crown apart, but there is a
tender point at the "collar," where they easily break
off short; with these also it is best to chop from below
or to use the chisel, making the cut well down in the
yellow rooty region. Veratrums divide much in the
same way, wanting a careful cut low down, the points
of their crowns being also very easy to break off. The
Christmas Rose is one of the most awkward plants to
divide successfully. It cannot be done in a hurry.
The only safe way is to wash the clumps well out
and look carefully for the points of attachment, and
cut them either with knife or chisel, according to their
position. In this case the chisel should be narrower
and sharper. Three-year-old tufts of St. Bruno's Lily
puzzled me at first. The rather fleshy roots are so
tightly interlaced that cutting is out of the question;
but I found out that if the tuft is held tight in the

two hands, and the hands are worked opposite ways
with a rotary motion of about a quarter of a circle,
that they soon come apart without being hurt in the
least.   Delphiniums easily break off at the crown if
they are broken up by hand, but the roots cut so easily
that it ought not to be a difficulty.

There are some plants in whose case one can never
be sure whether they will divide well or not, such as
Oriental Poppies and *Eryngium Olivieranum*.   They
behave in nearly the same way.   Sometimes a Poppy
or an Eryngium comes up with one thick root, impossible to divide, while the next door plant has a number
of roots that are ready to drop apart like a bunch of
Salsafy.

Everlasting Peas do nearly the same.   One may
dig up two plants—own brothers of say seven years
old—and a rare job it is, for they go straight down
into the earth nearly a yard deep.   One of them will
have a straight black post of a root $2\frac{1}{2}$ inches thick
without a break of any sort till it forks a foot underground, while the other will be a sort of loose rope of
separate roots from half to three-quarters of an inch
thick, that if carefully followed down and cleverly
dissected where they join, will make strong plants at
once.   But the usual way to get young plants of Everlasting Pea is to look out in earliest spring for the
many young growths that will be shooting, for these
if taken off with a good bit of the white underground
stem will root under a hand-light.

Most of the Primrose tribe divide pleasantly and easily : the worst are the *auricula* section ; with these, for outdoor planting, one often has to slice a main root down to give a share of root to the offset.

Where one is digging up plants with running roots, such as Gaultheria, Honeysuckle, Polygonum, Scotch Briars, and many of the *Rubus* tribe, or what is better, if one person is digging while another pulls up, it never does for the one who is pulling to give a steady haul ; this is sure to end in breakage, whereas a root comes up willingly and unharmed in loosened ground to a succession of firm but gentle tugs, and one soon learns to suit the weight of the pulls to the strength of the plant, and to learn its breaking strain.

Towards the end of October outdoor flowers in anything like quantity cannot be expected, and yet there are patches of bloom here and there in nearly every corner of the garden. The pretty Mediterranean Periwinkle (*Vinca acutiflora*) is in full bloom. As with many another southern plant that in its own home likes a cool and shady place, it prefers a sunny one in our latitude. The flowers are of a pale and delicate grey-blue colour, nearly as large as those of the common *Vinca major*, but they are borne more generously as to numbers on radical shoots that form thick, healthy-looking tufts of polished green foliage. It is not very common in gardens, but distinctly desirable.

In the bulb-beds the bright-yellow *Sternbergia lutea* is in flower. At first sight it looks something like a

Crocus of unusually firm and solid substance ; but it is an Amaryllis, and its pure and even yellow colouring is quite unlike that of any of the Crocuses. The numerous upright leaves are thick, deep green, and glossy. It flowers rather shyly in our poor soil, even in well-made beds, doing much better in chalky ground.

Czar Violets are giving their fine and fragrant flowers on stalks nine inches long. To have them at their best they must be carefully cultivated and liberally enriched. No plants answer better to good treatment, or spoil more quickly by neglect. A miserable sight is a forgotten violet-bed where they have run together into a tight mat, giving only few and poor flowers. I have seen the owner of such a bed stand over it and blame the plants, when he should have laid the lash on his own shoulders. Violets must be replanted every year. When the last rush of bloom in March is over, the plants are pulled to pieces, and strong single crowns from the outer edges of the clumps, or from the later runners, are replanted in good, well-manured soil, in such a place as will be somewhat shaded from summer sun. There should be eighteen inches between each plant, and as they make their growth, all runners should be cut off until August. They are encouraged by liberal doses of liquid manure from time to time, and watered in case of drought; and the heart of the careful gardener is warmed and gratified when friends, seeing them at midsummer, say (as has more than once happened), "What a nice batch of young Hollyhocks!"

In this simple matter of the culture of this good hardy Violet, my garden, though it is full of limitations, and in all ways falls short of any worthy ideal, enables me here and there to point out something that is worth doing, and to lay stress on the fact that the things worth doing are worth taking trouble about. But it is a curious thing that many people, even among those who profess to know something about gardening, when I show them something fairly successful—the crowning reward of much care and labour—refuse to believe that any pains have been taken about it. They will ascribe it to chance, to the goodness of my soil, and even more commonly to some supposed occult influence of my own—to anything rather than to the plain fact that I love it well enough to give it plenty of care and labour. They assume a tone of complimentary banter, kindly meant no doubt, but to me rather distasteful, to this effect: "Oh yes, of course it will grow for you; anything will grow for you; you have only to look at a thing and it will grow." I have to pump up a laboured smile and accept the remark with what grace I can, as a necessary civility to the stranger that is within my gates, but it seems to me evident that those who say these things do not understand the love of a garden.

I could not help rejoicing when such a visitor came to me one October. I had been saying how

necessary good and deep cultivation was, especially in
so very poor and shallow a soil as mine.   Passing up
through the copse where there were some tall stems
of *Lilium giganteum* bearing the great upturned pods
of seed, my visitor stopped and said, " I don't believe
a word about your poor soil—look at the growth of
that Lily.   Nothing could make that great stem ten
feet high in a poor soil, and there it is, just stuck
into the wood ! "   I said nothing, knowing that
presently I could show a better answer than I could
frame in words.   A little farther up in the copse we
came upon an excavation about twelve feet across
and four deep, and by its side a formidable mound
of sand, when my friend said, " Why are you making
all this mess in your pretty wood ? are you quarrying
stone, or is it for the cellar of a building ? and what
on earth are you going to do with that great heap of
sand ? why, there must be a dozen loads of it."   That
was my moment of secret triumph, but I hope I bore
it meekly as I answered, " I only wanted to plant a
few more of those big Lilies, and you see in my soil
they would not have a chance unless the ground was
thoroughly prepared; look at the edge of the scarp
and see how the solid yellow sand comes to within
four inches of the top; so I have a big wide hole dug;
and look, there is the donkey-cart coming with the
first load of Dahlia-tops and soft plants that have been
for the summer in the south border.   There will be
several of those little cartloads, each holding three

barrowfuls. As it comes into the hole, the men will chop it with the spade and tread it down close, mixing in a little sand. This will make a nice cool, moist bottom of slowly - rotting vegetable matter. Some more of the same kind of waste will come from the kitchen garden — cabbage-stumps, bean-haulm, soft weeds that have been hoed up, and all the greenest stuff from the rubbish-heap. Every layer will be chopped and pounded, and tramped down so that there should be as little sinking as possible afterwards. By this time the hole will be filled to within a foot of the top; and now we must get together some better stuff—road-scrapings and trimmings mixed with some older rubbish-heap mould, and for the top of all, some of our precious loam, and the soil of an old hotbed and some well-decayed manure, all well mixed, and then we are ready for the Lilies. They are planted only just underground, and then the whole bed has a surfacing of dead leaves, which helps to keep down weeds, and also looks right with the surrounding wild ground. The remains of the heap of sand we must deal with how we can; but there are hollows here and there in the roadway and paths, and a place that can be levelled up in the rubbish-yard, and some kitchen-garden paths that will bear raising, and so by degrees it is disposed of."

CHAPTER XII

NOVEMBER

Giant Christmas Rose—Hardy Chrysanthemums—Sheltering tender shrubs—Turfing by inoculation—Transplanting large trees— Sir Henry Steuart's experience early in the century—Collecting fallen leaves—Preparing grubbing tools—Butcher's Broom—Alexandrian Laurel—Hollies and Birches—A lesson in planting.

THE giant Christmas Rose (*Helleborus maximus*) is in full flower; it is earlier than the true Christmas Rose, being at its best by the middle of November. It is a large and massive flower, but compared with the later kinds has a rather coarse look. The bud and the back of the flower are rather heavily tinged with a dull pink, and it never has the pure-white colouring throughout of the later ones.

I have taken some pains to get together some really hardy November blooming Chrysanthemums. The best of all is a kind frequent in neighbouring cottage-gardens, and known hereabouts as Cottage Pink. I believe it is identical with Emperor of China, a very old sort that used to be frequent in greenhouse cultivation before it was supplanted by the many good kinds now grown. But its place is not indoors, but in

the open garden; if against a south or west wall, so
much the better. Perhaps one year in seven the bloom
may be spoilt by such a severe frost as that of October
1895, but it will bear unharmed several degrees of
frost and much rain. I know no Chrysanthemum of
so true a pink colour, the colour deepening to almost
crimson in the centre. After the first frost the foliage
of this kind turns to a splendid colour, the green of
the leaves giving place to a rich crimson that sometimes clouds the outer portion of the leaf, and often
covers its whole expanse. The stiff, wholesome foliage
adds much to the beauty of the outdoor kinds, contrasting most agreeably with the limp, mildewed leafage
of those indoors. Following Cottage Pink is a fine
pompone called Soleil d'Or, in colour the richest deep
orange, with a still deeper and richer coloured centre.
The beautiful crimson Julie Lagravère flowers at the
same time. Both are nearly frost-proof, and true hardy
November flowers.

The first really frosty day we go to the upper part
of the wood and cut out from among the many young
Scotch Firs as many as we think will be wanted for
sheltering plants and shrubs of doubtful hardiness.
One section of the high wall at the back of the flower
border is planted with rather tender things, so that the
whole is covered with sheltering fir-boughs. Here are
Loquat, Fuchsia, Pomegranate, *Edwardsia*, *Piptanthus*,
and *Choisya*, and in the narrow border at the foot of
the wall, *Crinum*, *Nandina*, *Clerodendron*, and *Hydrangea*.

K

In the broad border in front of the wall nothing needs protection except Tritomas; these have cones of coal-ashes heaped over each plant or clump. The Crinums also have a few inches of ashes over them.

Some large Hydrangeas in tubs are moved to a sheltered place and put close together, a mound of sand being shovelled up all round to nearly the depth of the tubs; then a wall is made of thatched hurdles, and dry fern is packed well in among the heads of the plants. They would be better in a frost-proof shed, but we have no such place to spare.

The making of a lawn is a difficulty in our very poor sandy soil. In this rather thickly-populated country the lords of the manor had been so much pestered for grants of road-side turf, and the privilege when formerly given had been so much abused, that they have agreed together to refuse all applications. Opportunities of buying good turf do not often occur, and sowing is slow, and not satisfactory. I am told by a seedsman of the highest character that it is almost impossible to get grass seed clean and true to name from the ordinary sources; the leading men therefore have to grow their own.

In my own case, having some acres of rough heath and copse where the wild grasses are of fine-leaved kinds, I made the lawn by inoculation. The ground was trenched and levelled, then well trodden and raked, and the surface stones collected. Tufts of the wild grass were then forked up, and were pulled into pieces

about the size of the palm of one's hand, and laid down
eight inches apart, and well rolled in. During the
following summer we collected seed of the same grasses
to sow early in spring in any patchy or bare places.
One year after planting the patches had spread to
double their size, and by the second year had nearly
joined together. The grasses were of two kinds only,
namely, Sheep's Fescue (*Festuca ovina*) and Crested
Dog's-tail (*Agrostis canina*). They make a lawn of
a quiet, low-toned colour, never of the bright green of
the rather coarser grasses; but in this case I much
prefer it; it goes better with the Heath and Fir and
Bracken that belong to the place. In point of labour,
a lawn made of these fine grasses has the great merit
of only wanting mowing once in three weeks.

I have never undertaken the transplanting of large
trees, but there is no doubt that it may be done with
success, and in laying out a new place where the site
is bare, if suitable trees are to be had, it is a plan
much to be recommended. It has often been done
of late years, but until a friend drew my attention to
an article in the *Quarterly Review*, dated March 1828,
I had no idea that it had been practised on a large
scale so early in the century. The article in question
was a review of " The Planter's Guide," by Sir Henry
Steuart, Bart., LL.D. (Edinburgh, 1828.) It quoted
the opinion and observation of a committee of gentle-
men, among whom was Sir Walter Scott, who visited

Allanton (Sir Henry Steuart's place) in September
1828, when the trees had been some years planted.
They found them growing "with vigour and luxuriance,
and in the most exposed situations making shoots of
eighteen inches. . . . From the facts which they wit-
nessed the committee reported it as their unanimous
opinion that the art of transplantation, as practised by
Sir Henry Steuart, is calculated to accelerate in an
extraordinary degree the power of raising wood, whether
for beauty or shelter."

The reviewer then quotes the method of transplantation, describing the extreme care with which
the roots are preserved, men with picks carefully
trying round the ground beneath the outer circumference of the branches for the most outlying rootlets,
and then gradually approaching the bole. The greatest
care was taken not to injure any root or fibre, these
as they were released from the earth being tied up,
and finally the transplanting machine, consisting of a
strong pole mounted on high wheels, was brought close
to the trunk and attached to it, and the tree when
lowered, carefully transported to its new home. Every
layer of roots was then replanted with the utmost care,
with delicate fingering and just sufficient ramming,
and in the end the tree stood without any artificial
support whatever, and in positions exposed to the
fiercest gales.

The average size of tree dealt with seems to have
had a trunk about a foot in diameter, but some were

removed with complete success whose trunks were two feet thick. In order that his trees might be the better balanced in shape, Sir Henry boldly departed from the older custom of replanting a tree in its original aspect, for he reversed the aspect, so that the more stunted and shorter-twigged weather side now became the lee side, and could grow more freely.

He insists strongly on the wisdom of transplanting only well-weathered trees, and not those of tender constitution that had been sheltered by standing among other close growths, pointing out that these have a tenderer bark and taller top and roots less well able to bear the strain of wind and weather in the open.

He reckons that a transplanted tree is in full new growth by the fourth or fifth year, and that an advantage equal to from thirty to forty years' growth is gained by the system. As for the expense of the work, Sir Henry estimated that his largest trees each cost from ten to thirteen shillings to take up, remove half a mile, and replant. In the case of large trees the ground that was to receive them was prepared a twelvemonth beforehand.

Now, in the third week of November, the most pressing work is the collecting of leaves for mulching and leaf-mould. The oaks have been late in shedding their leaves, and we have been waiting till they are down. Oak-leaves are the best, then hazel, elm, and Spanish chestnut. Birch and beech are not so good; beech-leaves especially take much too long to decay. This is, no doubt, the reason why nothing grows willingly under beeches.   Horse and cart and three hands go out into the lanes for two or three days, and the loads that come home go three feet deep into the bottom of a range of pits.   The leaves are trodden down close and covered with a layer of mould, in which winter salad stuff is immediately planted.   The mass of leaves will soon begin to heat, and will give a pleasant bottom-heat throughout the winter.   Other loads of leaves go into an open pen about ten feet square and five feet deep. Two such pens, made of stout oak post and rail and upright slabs, stand side by side in the garden yard. The one newly filled has just been emptied of its two-year-old leaf-mould, which has gone as a nourishing and protecting mulch over beds of Daffodils and choice bulbs and Alströmerias, some being put aside in reserve for potting and various uses.   The other pen remains full of the leaves of last year, slowly rotting into whole-some plant-food.

With works of wood-cutting and stump-grubbing near at hand, we look over the tools and see that all are in readiness for winter work.   Axes and hand-bills are ground, fag-hooks sharpened, picks and mattocks sent to the smithy to be drawn out, the big cross-cut saw fresh sharpened and set, and the hand-saws and frame-saws got ready.   The rings of the bittle are tightened and wedged up, so that its heavy head may not split when the mighty blows, flung into the tool

PENS FOR STORING DEAD LEAVES.

CAREFUL WILD-GARDENING—WHITE FOXGLOVES AT THE EDGE OF
THE FIR WOOD. (*See page* 270.)

with a man's full strength, fall on the heads of the great iron wedges.

Some thinning of birch-trees has to be done in the lowest part of the copse, not far from the house. They are rather evenly distributed on the ground, and I wish to get them into groups by cutting away superfluous trees. On the neighbouring moorland and heathy uplands they are apt to grow naturally in groups, the individual trees generally bending outward towards the free, open space, the whole group taking a form that is graceful and highly pictorial. I hope to be able to cut out trees so as to leave the remainder standing in some such way. But as a tree once cut cannot be put up again, the condemned ones are marked with bands of white paper right round the trunks, so that they can be observed from all sides, thus to give a chance of reprieve to any tree that from any point of view may have pictorial value.

Frequent in some woody districts in the south of England, though local, is the Butcher's Broom (*Ruscus aculeatus*). Its stiff green branches that rise straight from the root bear small, hard leaves, armed with a sharp spine at the end. The flower, which comes in early summer, is seated without stalk in the middle of the leaf, and is followed by a large red berry. In country places where it abounds, butchers use the twigs tied in bunches to brush the little chips of meat off their great chopping-blocks, that are made of solid sections of elm trees, standing three and a half feet high and about two and a half feet across. Its beautiful garden relative, the Alexandrian or Victory Laurel (*Ruscus racemosus*), is also now just at its best. Nothing makes a more beautiful wreath than two of its branches, suitably arched and simply bound together near the butts and free ends. It is not a laurel, but a *Ruscus*, the name laurel having probably grown on to it by old association with any evergreen suitable for a victor's wreath. It is a slow-growing plant, but in time makes handsome tufts of its graceful branches. Few plants are more exquisitely modelled, to use a term familiar to the world of fine art, or give an effect of more delicate and perfect finish. It is a valuable plant in a shady place in good, cool soil. Early in summer, when the young growths appear, the old, then turning rusty, should be cut away.

No trees group together more beautifully than Hollies and Birches. One such happy mixture in one part of the copse suggested further plantings of Holly, Birches being already in abundance. Every year some more Hollies are planted; those put in nine years ago are now fifteen feet high, and are increasing fast. They are slow to begin growth after transplanting, perhaps because in our very light soil they cannot be moved with a "ball"; all the soil shakes away, and leaves the root naked; but after about three years, when the roots have got good hold and begun to ramble, they grow away well. The trunk of an old Holly has a smooth pale-grey bark, and sometimes

HOLLY STEMS IN AN OLD HEDGE-ROW.

a slight twist, that makes it look like the gigantic
bone of some old-world monster. The leaves of some
old trees, especially if growing in shade, change their
shape, losing the side prickles and becoming longer
and nearly flat and more of a dark bottle-green colour,
while the lower branches and twigs, leafless except
towards their ends, droop down in a graceful line that
rises again a little at the tip.

The leaves are all down by the last week of
November, and woodland assumes its winter aspect;
perhaps one ought rather to say, some one of its
infinite variety of aspects, for those who live in such
country know how many are the winter moods of
forest land, and how endless are its variations of
atmospheric effect and pictorial beauty—variations
much greater and more numerous than are possible
in summer.

With the wind in the south-west and soft rain
about, the twigs of the birches look almost crimson,
while the dead bracken at their foot, half-draggled
and sodden with wet, is of a strong, dark rust colour.
Now one sees the full value of the good evergreens,
and, rambling through woodland, more especially of
the Holly, whether in bush or tree form, with its
masses of strong green colour, dark and yet never
gloomy. Whether it is the high polish of the leaves,
or the lively look of their wavy edges, with the short
prickles set alternately up and down, or the brave way
the tree has of shooting up among other thick growth

or its massive sturdiness on a bare hillside, one cannot say, but a Holly in early winter, even without berries, is always a cheering sight. John Evelyn is eloquent in his praise of this grand evergreen, and lays special emphasis on this quality of cheerfulness.

Near my home is a little wild valley, whose planting, wholly done by Nature, I have all my life regarded with the most reverent admiration.

The arable fields of an upland farm give place to hazel copses as the ground rises. Through one of these a deep narrow lane, cool and dusky in summer from its high steep banks and over-arching foliage, leads by a rather sudden turn into the lower end of the little valley. Its grassy bottom is only a few yards wide, and its sides rise steeply right and left. Looking upward through groups of wild bushes and small trees, one sees thickly-wooded ground on the higher levels. The soil is of the very poorest ; ridges of pure yellow sand are at the mouths of the many rabbit-burrows. The grass is of the short fine kinds of the heathy uplands. Bracken grows low, only from one to two feet high, giving evidence of the poverty of the soil, and yet it seems able to grow in perfect beauty clumps of Juniper and Thorn and Holly, and Scotch Fir on the higher ground.

On the steeply-rising banks are large groups of Juniper, some tall, some spreading, some laced and wreathed about with tangles of Honeysuckle, now in brown winter dress, and there are a few bushes of

WILD JUNIPERS.

Spindle-tree, whose green stems and twigs look strangely green in winter. The Thorns stand some singly, some in close companionship, impenetrable masses of short-twigged prickly growth, with here and there a wild Rose shooting straight up through the crowded branches. One thinks how lovely it will be in early June, when the pink Rose-wreaths are tossing out of the foamy sea of white Thorn blossom. The Hollies are towering masses of health and vigour. Some of the groups of Thorn and Holly are intermingled; all show beautiful arrangements of form and colour, such as are never seen in planted places. The track in the narrow valley trends steadily upwards and bears a little to the right. High up on the left-hand side is an old wood of Scotch Fir. A few detached trees come half-way down the valley bank to meet the gnarled, moss-grown Thorns and the silver-green Junipers. As the way rises some Birches come in sight, also at home in the sandy soil. Their graceful, lissome spray moving to the wind looks active among the stiffer trees, and their white stems shine out in startling contrast to the other dusky foliage. So the narrow track leads on, showing the same kinds of tree and bush in endless variety of beautiful grouping, under the sombre half-light of the winter day. It is afternoon, and as one mounts higher a pale bar of yellow light gleams between the farther tree-stems, but all above is grey, with angry, blackish drifts of ragged wrack. Now the valley opens out to a nearly level space of rough

grass, with grey tufts that will be pink bell-heather
in summer, and upstanding clumps of sedge that tell
of boggy places. In front and to the right are dense
fir-woods. To the left is broken ground and a steep-
sided hill, towards whose shoulder the track rises.
Here are still the same kinds of trees, but on the
open hillside they have quite a different effect. Now
I look into the ruddy heads of the Thorns, bark and
fruit both of rich warm colouring, and into the upper
masses of the Hollies, also reddening into wealth of
berry.

Throughout the walk, pacing slowly but steadily
for nearly an hour, only these few kinds of trees have
been seen, Juniper, Holly, Thorn, Scotch Fir, and Birch
(a few small Oaks excepted), and yet there has not
been once the least feeling of monotony, nor, returning
downward by the same path, could one wish anything
to be altered or suppressed or differently grouped. And
I have always had the same feeling about any quite
wild stretch of forest land. Such a bit of wild forest
as this small valley and the hilly land beyond are
precious lessons in the best kind of tree and shrub
planting. No artificial planting can ever equal that
of Nature, but one may learn from it the great lesson
of the importance of moderation and reserve, of sim-
plicity of intention, and directness of purpose, and the
inestimable value of the quality called " breadth " in
painting. For planting ground is painting a landscape with living things; and as I hold that good

WILD JUNIPERS.

gardening takes rank within the bounds of the fine arts, so I hold that to plant well needs an artist of no mean capacity. And his difficulties are not slight ones, for his living picture must be right from all points, and in all lights.

No doubt the planting of a large space with a limited number of kinds of trees cannot be trusted to all hands, for in those of a person without taste or the more finely-trained perceptions the result would be very likely dull or even absurd. It is not the paint that make the picture, but the brain and heart and hand of the man who uses it.

CHAPTER XIII

DECEMBER

The woodman at work—Tree-cutting in frosty weather—Preparing
sticks and stakes—Winter Jasmine—Ferns in the wood-walk
—Winter colour of evergreen shrubs—Copse-cutting—Hoop-
making—Tools used—Sizes of hoops—Men camping out—
Thatching with hoop-chips—The old thatcher's bill.

It is good to watch a clever woodman and see how
much he can do with his simple tools, and how easily
one man alone can deal with heavy pieces of timber. An
oak trunk, two feet or more thick, and weighing perhaps
a ton, lies on the ground, the branches being already
cut off. He has to cleave it into four, and to remove it
to the side of a lane one hundred feet away. His tools
are an axe and one iron wedge. The first step is the
most difficult—to cut such a nick in the sawn surface
of the butt of the trunk as will enable the wedge to
stick in. He holds the wedge to the cut and hammers
it gently with the back of the axe till it just holds,
then he tries a moderate blow, and is quite prepared
for what is almost sure to happen—the wedge springs
out backwards; very likely it springs out for three
or four trials, but at last the wedge bites and he
can give it the dexterous, rightly-placed blows that

THE WOODMAN.

slowly drive it in. Before the wedge is in half its length a creaking sound is heard; the fibres are beginning to tear, and a narrow rift shows on each side of the iron. A few more strokes and the sound of the rending fibres is louder and more continuous, with sudden cracking noises, that tell of the parting of larger bundles of fibres, that had held together till the tremendous rending power of the wedge at last burst them asunder. Now the man looks out a bit of strong branch about four inches thick, and with the tree-trunk as a block and the axe held short in one hand as a chopper, he makes a wooden wedge about twice the size of the iron one, and drives it into one of the openings at its side. For if you have only one iron wedge, and you drive it tight into your work, you can neither send it farther nor get it out, and you feel and look foolish. The wooden wedge driven in releases the iron one, which is sent in afresh against the side of the wedge of oak, the trunk meanwhile rending slowly apart with much grieving and complaining of the tearing fibres. As the rent opens the axe cuts across diagonal bundles of fibres that still hold tightly across the widening rift. And so the work goes on, the man unconsciously exercising his knowledge of his craft in placing and driving the wedges, the helpless wood groaning and creaking and finally falling apart as the last holding fibres are severed by the axe. Meanwhile the raw green wood gives off a delicious scent, sweet and sharp and refreshing, not unlike the smell of apples crushing in the cider-press.

The woodman has still to rend the two halves of the trunk, but the work is not so heavy and goes more quickly. Now he has to shift them to the side of the rough track that serves as a road through the wood. They are so heavy that two men could barely lift them, and he is alone. He could move them with a lever, that he could cut out of a straight young tree, a foot or so at a time at each end, but it is a slow and clumsy way; besides, the wood is too much encumbered with undergrowth. So he cuts two short pieces from a straight bit of branch four inches or five inches thick, levers one of his heavy pieces so that one end points to the roadway, prises up this end and kicks one of his short pieces under it close to the end, settling it at right angles with gentle kicks. The other short piece is arranged in the same way, a little way beyond the middle of the length of quartered trunk. Now, standing behind it, he can run the length easily along on the two rollers, till the one nearest him is left behind; this one is then put under the front end of the weight, and so on till the road is reached.

Trees that stand where paths are to come, or that for any reason have to be removed, root and all, are not felled with axe or saw, but are grubbed down. The earth is dug away next to the tree, gradually exposing the roots; these are cut through

GRUBBING A TREE-STUMP.

FELLING AND GRUBBING TOOLS. (*See page* 150.)

with axe or mattock close to the butt, and again
about eighteen inches away, so that by degrees a
deep trench, eighteen inches wide, is excavated round
the butt. A rope is fastened at the right distance
up the trunk, when, if the tree does not hold by a
very strong tap-root, a succession of steady pulls will
bring it down; the weight of the top thus helping
to prise the heavy butt out of the ground. We come
upon many old stumps of Scotch fir, the remains
of the original wood; they make capital firewood,
though some burn rather too fiercely, being full of
turpentine. Many are still quite sound, though it
must be six-and-twenty years since they were felled.
They are very hard to grub, with their thick tap-
roots and far-reaching laterals, and still tougher to
split up, their fibres are so much twisted, and the
dark-red heart-wood has become hardened till it
rings to a blow almost like metal. But some, whose
roots have rotted, come up more easily, and with
very little digging may be levered out of the ground
with a long iron stone-bar, such as they use in the
neighbouring quarries, putting the point of the bar
under the "stam," and having a log of wood for a
hard fulcrum. Or a stout young stem of oak or
chestnut is used for a lever, passing a chain under
the stump and over the middle of the bar and prising
upwards with the lever. "Stam" is the word always
used by the men for any stump of a tree left in the
ground.

L

A spell of frosty days at the end of December puts a stop to all planting and ground work. Now we go into the copse and cut the trees that have been provisionally marked, judged, and condemned, with the object of leaving the remainder standing in graceful groups. The men wonder why I cut some of the trees that are best and straightest and have good tops, and leave those with leaning stems. Anything of seven inches or less diameter is felled with the axe, but thicker trees with the cross-cut saw. For these our most active fellow climbs up the tree with a rope, and makes it fast to the trunk a good way up, then two of them, kneeling, work the saw. When it has cut a third of the way through, the rope is pulled on the side opposite the cut to keep it open and let the saw work free. When still larger trees are sawn down this is done by driving in a wedge behind the saw, when the width of the saw-blade is rather more than buried in the tree. When the trunk is nearly sawn through, it wants care and judgment to see that the saw does not get pinched by the weight of the tree; the clumsy workman who fails to clear his saw gets laughed at, and probably damages his tool. Good straight trunks of oak and chestnut are put aside for special uses; the rest of the larger stuff is cut into cordwood lengths of four feet. The heaviest of these are split up into four pieces to make them easier to load and carry away, and eventually to saw up into firewood.

The best of the birch tops are cut into pea-sticks, a clever, slanting cut with the hand bill leaving them pointed and ready for use. Throughout the copse are "stools" of Spanish chestnut, cut about once in five years. From this we get good straight stakes for Dahlias and Hollyhocks, also bean-poles; while the rather straight-branched boughs are cut into branching sticks for Michaelmas Daisies, and special lengths are got ready for various kinds of plants—Chrysanthemums, Lilies, Pæonies, and so on. To provide all this in winter, when other work is slack or impossible, is an important matter in the economy of a garden, for all gardeners know how distressing and harassing it is to find themselves without the right sort of sticks or stakes in summer, and what a long job it then seems to have to look them up and cut them, of indifferent quality, out of dry faggots. By the plan of preparing all in winter no precious time is lost, and a tidy withe-bound bundle of the right sort is always at hand. The rest of the rough spray and small branching stuff is made up into faggots to be chopped up for fire-lighting; the country folk still use the old word "bavin" for faggots. The middle-sized branches—anything between two inches and six inches in diameter—are what the woodmen call "top and lop"; these are also cut into convenient lengths, and are stacked in the barn, to be cut into billets for next year's fires in any wet or frosty weather, when outdoor work is at a standstill.

What a precious winter flower is the yellow Jasmine (*Jasminum nudiflorum*). Though hard frost spoils the flowers then expanded, as soon as milder days come the hosts of buds that are awaiting them burst into bloom. Its growth is so free and rapid that one has no scruple about cutting it freely; and great branching sprays, cut a yard or more long, arranged with branches of Alexandrian Laurel or other suitable foliage—such as Andromeda or Gaultheria—are beautiful as room decoration.

Christmas Roses keep on flowering bravely, in spite of our light soil and frequent summer drought, both being unfavourable conditions; but bravest of all is the blue Algerian Iris (*Iris stylosa*), flowering freely as it does, at the foot of a west wall, in all open weather from November till April.

In the rock-garden at the edge of the copse the creeping evergreen *Polygala chamœbuxus* is quite at home in beds of peat among mossy boulders. Where it has the ground to itself, this neat little shrub makes close tufts only four inches or five inches high, its wiry branches being closely set with neat, dark-green, box-like leaves; though where it has to struggle for life among other low shrubs, as may often be seen in the Alps, the branches elongate, and will run bare for two feet or three feet to get the leafy end to the light. Even now it is thickly set with buds and has a few expanded flowers. This bit of rock-garden is mostly planted with dwarf shrubs — *Skimmia*, Bog-myrtle,

Alpine Rhododendrons, *Gaultheria*, and *Andromeda*, with
drifts of hardy ferns between, and only a few "soft"
plants. But of these, two are now conspicuously
noticeable for foliage—the hardy Cyclamens and the
blue Himalayan Poppy (*Meconopsis Wallichi*). Every
winter I notice how bravely the pale woolly foliage of
this plant bears up against the early winter's frost
and wet.

The wood-walk, whose sloping banks are planted
with hardy ferns in large groups, shows how many of
our common kinds are good plants for the first half
of the winter. Now, only a week before Christmas, the
male fern is still in handsome green masses; *Blechnum*
is still good, and common Polypody at its best. The
noble fronds of the Dilated Shield-fern are still in
fairly good order, and *Ceterach* in rocky chinks is in
fullest beauty. Beyond, in large groups, are prosper-
ous-looking tufts of the Wood-rush (*Luzula sylvatica*);
then there is wood as far as one can see, here mostly
of the silver-stemmed Birch and rich-green Holly, with
the woodland carpet of dusky low-toned bramble and
quiet dead leaf and brilliant moss.

By the middle of December many of the evergreen
shrubs that thrive in peat are in full beauty of foliage.
Andromeda Catesbœi is richly coloured with crimson
clouds and splashes; Skimmias are at their best and
freshest, their bright, light green, leathery foliage defy-
ing all rigours of temperature or weather. Pernettyas
are clad in their strongest and deepest green leafage,

and show a richness and depth of colour only surpassed by that of the yew hedges.

Copse-cutting is one of the harvests of the year for labouring men, and all the more profitable that it can go on through frosty weather. A handy man can earn good wages at piece-work, and better still if he can cleave and shave hoops. Hoop-making is quite a large industry in these parts, employing many men from Michaelmas to March. They are barrel-hoops, made of straight poles of six years' growth. The wood used is Birch, Ash, Hazel and Spanish Chestnut. Hazel is the best, or as my friend in the business says, " Hazel, that's the master !" The growths of the copses are sold by auction in some near county town, as they stand, the buyer clearing them during the winter. They are cut every six years, and a good copse of Chestnut has been known to fetch £54 an acre.

A good hoop-maker can earn from twenty to twenty-five shillings a week. He sets up his brake, while his mate, who will cleave the rods, cuts a post about three inches thick, and fixes it into the ground so that it stands about three feet high. To steady it he drives in another of rather curly shape by its side, so that the tops of the two are nearly even, but the foot of the curved spur is some nine inches away at the bottom, with its top pressing hard against the upright. To stiffen it still more he makes a long withe of a straight hazel rod, which he twists into a rope by holding the butt tightly under his left foot and

HOOF-MAKING IN THE WOODS.

twisting with both hands till the fibres are wrenched open and the withe is ready to spring back and wind upon itself. With this he binds his two posts together, so that they stand perfectly rigid. On this he cleaves the poles, beginning at the butt. The tool is a small one-handed adze with a handle like a hammer. A rod is usually cleft in two, so that it is only shaved on one side; but sometimes a pole of Chestnut, a very quick-growing wood, is large enough to cleave into eight, and when the wood is very clean and straight they can sometimes get two lengths of fourteen feet out of a pole.

The brake is a strong flat-shaped post of oak set up in the ground to lean a little away from the workman. It stands five and a half feet out of the ground. A few inches from its upper end it has a shoulder cut in it which acts as the fulcrum for the cross-bar that supports the pole to be shaved, and that leans down towards the man. The relative position of the two parts of the brake reminds one of the mast and yard of a lateen-rigged boat. The bar is nicely balanced by having a hazel withe bound round a groove at its upper short end, about a foot beyond the fulcrum, while the other end of the withe is tied round a heavy bit of log or stump that hangs clear of the ground and just balances the bar, so that it see-saws easily. The cleft rod that is to be shaved lies along the bar, and an iron pin that passes through the head of the brake just above the point where the bar rides over its

ERRATUM.

Page 167, line 5, *for* " beginning at the butt,"
read " beginning at the top."

shoulder, nips the hoop as the weight of the stroke comes upon it; the least lifting of the bar releases the hoop, which is quickly shifted onwards for a new stroke. The shaving tool is a strong two-handled draw-knife, much like the tool used by wheelwrights. It is hard work, "wunnerful tryin' across the chest."

The hoops are in several standard lengths, from fourteen to two and a half feet. The longest go to the West Indies for sugar hogsheads, and some of the next are for tacking round pipes of wine. The wine is in well-made iron-hooped barrels, but the wooden hoops are added to protect them from the jarring and bumping when rolled on board ship, and generally to save them during storage and transit. These hoops are in two sizes, called large and small pipes. A thirteen-foot size go to foreign countries for training vines on. A large quantity that measure five feet six inches, and called " long pinks," are for cement barrels. A length of seven feet six inches are used for herring barrels, and are called kilderkins, after the name of the size of tub. Smaller sizes go for gunpowder barrels, and for tacking round packing-cases and tea-chests.

The men want to make all the time they can in the short winter daylight, and often the work is some miles from home, so if the weather is not very cold they make huts of the bundles of rods and chips, and sleep out on the job. I always admire the neatness with which the bundles are fastened up, and the strength of the withe-rope that binds them, for sixty

Shed-roof, thatched with Hoop-chip.

Hoop-shaving.

hoops, or thirty pairs, as they call them, of fourteen feet, are a great weight to be kept together by four slight hazel bands.

In this industry there is a useful by-product in the shavings or chips, as they call them. They are eighteen inches to two feet long, and are made up into small faggots or bundles and stacked up for six months to a year to dry, and then sell readily at two-pence a bundle to cut up for fire-lighting. They also make a capital thatch for sheds, a thatch nearly a foot thick, warm in winter, and cool in summer, and durable, for if well made it will last for forty years. I got a clever old thatcher to make me a hoop-chip roof for the garden shed; it was a long job, and he took his time (although it was piece-work), preparing and placing each handful of chips as carefully as if he was making a wedding bouquet. He was one of the old sort—no scamping of work for him; his work was as good as he could make it, and it was his pride and delight. The roof was prepared with strong laths nailed horizontally across the rafters as if for tiling, but farther apart; and the chips, after a number of handfuls had been duly placed and carefully poked and patted into shape, were bound down to the laths with soft tarred cord guided by an immense iron needle. The thatching, as in all cases of roof-covering, begins at the eaves, so that each following layer laps over the last. Only the ridge has to be of straw, because straw can be bent over; the chips are too rigid. When the thatch is all in place the whole is "drove," that is, beaten up close with a wooden bat that strikes against the ends of the chips and drives them up close, jamming them tight into the fastening. After six months of drying summer weather he came and drove it all over again.

Thatching is done by piece-work, and paid at so much a "square" of ten by ten feet. When I asked for his bill, the old man brought it made out on a hazel stick, in a manner either traditional, or of his own devising. This is how it runs, in notches about half an inch long, and dots dug with the point of the knife. It means, "To so much work done, £4, 5s. 0d."

ΙΙXXX·Ι·, ΙΙXXXX·ΙΙΛ ΙΙΙΙΛXX,ΙΙXX

CHAPTER XIV

LARGE AND SMALL GARDENS

A well done villa garden—A small town garden—Two delightful
gardens of small size—Twenty acres within the walls—A large
country house and its garden—Terrace—Lawn—Parterre—
Free garden—Kitchen garden—Buildings—Ornamental orchard
—Instructive mixed gardens—Mr. Wilson's at Wisley—A
window garden.

THE size of a garden has very little to do with its
merit. It is merely an accident relating to the cir-
cumstances of the owner. It is the size of his heart
and brain and goodwill that will make his garden
either delightful or dull, as the case may be, and either
leave it at the usual monotonous dead-level, or raise it,
in whatever degree he may, towards that of a work of
fine art. If a man knows much, it is more difficult
for him to deal with a small space than a larger, for
he will have to make the more sacrifice; but if he is
wise he will at once make up his mind about what he
will let go, and how he may best treat the restricted
space. Some years ago I visited a small garden
attached to a villa on the outskirts of a watering-place
on the south coast. In ordinary hands it would have
been a perfectly commonplace thing, with the usual

weary mixture, and exhibiting the usual distressing symptoms that come in the train of the ministrations of the jobbing-gardener. In size it may have been a third of an acre, and it was one of the most interesting and enjoyable gardens I have ever seen, its master and mistress giving it daily care and devotion, and enjoying to the full its glad response of grateful growth. The master had built with his own hands, on one side where more privacy was wanted, high rugged walls, with spaces for many rock-loving plants, and had made the wall die away so cleverly into the rock-garden, that the whole thing looked like a garden founded on some ancient ruined structure. And it was all done with so much taste that there was nothing jarring or strained-looking, still less anything cock-neyfied, but all easy and pleasant and pretty, while the happy look of the plants at once proclaimed his sympathy with them, and his comprehensive knowledge of their wants. In the same garden was a walled enclosure where Tree Pæonies and some of the hardier of the oriental Rhododendrons were thriving, and there were pretty spaces of lawn, and flower border, and shrub clump, alike beautiful and enjoyable, all within a small space, and yet not crowded—the garden of one who was a keen flower lover, as well as a world-known botanist.

I am always thankful to have seen this garden, because it showed me, in a way that had never been so clearly brought home to me, how much may be done in a small space.

Another and much smaller garden that I remember with pleasure was in a sort of yard among houses, in a country town. The house it belonged to, a rather high one, was on its east side, and halfway along on the south; the rest was bounded by a wall about ten feet high. Opposite the house the owner had built of rough blocks of sandstone what served as a workshop, about twelve feet long, along the wall, and six feet wide within. A low archway of the same rough stone was the entrance, and immediately above it a lean-to roof sloped up to the top of the wall, which just here had been carried a little higher. The roof was of large flat sandstones, only slightly lapping over each other, with spaces and chinks where grew luxuriant masses of Polypody Fern. It was contrived with a cement bed, so that it was quite weather-tight, and the room was lighted by a skylight at one end that did not show from the garden. A small surface of lead-flat, on a level with the top of the wall, in one of the opposite angles, carried an old oil-jar, from which fell masses of gorgeous Tropæolum, and the actual surface of the flat was a garden of Stonecrops. The rounded coping of the walls, and the joints in many places (for the wall was an old one), were gay with yellow Corydalis and Snapdragons and more Stonecrops. The little garden had a few pleasant flowering bushes, Ribes and Laurustinus, a Bay and an Almond tree. In the coolest and shadiest corner were a fern-grotto and a tiny tank. The rest of the garden,

only a few yards across, was laid out with a square
bed in the middle, and a little path round, then a
three-feet-wide border next the wall, all edged with
rather tall-grown Box.   The middle bed had garden
Roses and Carnations, and Mignonette and Stocks.
All round were well-chosen plants and shrubs, looking
well and happy, though in a confined and rather airless
space.   Every square foot had been made the most
of with the utmost ingenuity, but the ingenuity was
always directed by good taste, so that nothing looked
crowded or out of place.

And I think of two other gardens of restricted
space, both long strips of ground walled at the sides,
whose owners I am thankful to count among my
friends — one in the favoured climate of the Isle
of Wight, a little garden where I suppose there are
more rare and beautiful plants brought together within
a small space than perhaps in any other garden of
the same size in England; the other in a cathedral
town, now a memory only, for the master of what was
one of the most beautiful gardens I have ever seen
now lives elsewhere.   The garden was long in shape,
and divided about midway by a wall.   The division
next the house was a quiet lawn, with a mulberry
tree and a few mounded borders near the sides that
were unobstrusive, and in no way spoilt the quiet
feeling of the lawn space.   Then a doorway in the
dividing wall led to a straight path with a double
flower border.   I suppose there was a vegetable garden

behind the borders, but of that I have no recollection, only a vivid remembrance of that brilliantly beautiful mass of flowers. The picture was good enough as one went along, especially as at the end one came first within sound and then within sight of a rushing river, one of those swift, clear, shallow streams with stony bottom that the trout love; but it was ten times more beautiful on turning to go back, for there was the mass of flowers, and towering high above it the noble mass of the giant structure—one of the greatest and yet most graceful buildings that has ever been raised by man to the glory of God.

It is true that it is not every one that has the advantage of a garden bounded by a river and a noble church, but even these advantages might have been lost by vulgar or unsuitable treatment of the garden. But the mind of the master was so entirely in sympathy with the place, that no one that had the privilege of seeing it could feel that it was otherwise than right and beautiful.

Both these were the gardens of clergymen; indeed, some of our greatest gardeners are, and have been, within the ranks of the Church. For have we not a brilliantly-gifted dignitary whose loving praise of the Queen of flowers has become a classic? and have we not among churchmen the greatest grower of seedling Daffodils the world has yet seen, and other names of clergymen honourably associated with Roses and Auriculas and Tulips and other good flowers, and all

greatly to their bettering ? The conditions of the life
of a parish priest would tend to make him a good
gardener, for, while other men roam about, he stays
mostly at home, and to live with one's garden is one
of the best ways to ensure its welfare. And then,
among the many anxieties and vexations and dis-
appointments that must needs grieve the heart of
the pastor of his people, his garden, with its wholesome labour and all its lessons of patience and trust
and hopefulness, and its comforting power of solace,
must be one of the best of medicines for the healing
of his often sorrowing soul.

I do not envy the owners of very large gardens.
The garden should fit its master or his tastes just
as his clothes do ; it should be neither too large nor
too small, but just comfortable. If the garden is
larger than he can individually govern and plan and
look after, then he is no longer its master but its slave,
just as surely as the much-too-rich man is the slave
and not the master of his superfluous wealth. And
when I hear of the great place with a kitchen garden
of twenty acres within the walls, my heart sinks as I
think of the uncomfortable disproportion between the
man and those immediately around him, and his vast
output of edible vegetation, and I fall to wondering
how much of it goes as it should go, or whether the
greater part of it does not go dribbling away, leaking
into unholy back-channels ; and of how the looking
after it must needs be subdivided ; and of how many

side-interests are likely to steal in, and altogether how great a burden of anxiety or matter of temptation it must give rise to. A grand truth is in the old farmer's saying, "The master's eye makes the pig fat;" but how can any one master's eye fat that vast pig of twenty acres, with all its minute and costly cultivation, its two or three crops a year off all ground given to soft vegetables, its stoves, greenhouses, orchid and orchard houses, its vineries, pineries, figgeries, and all manner of glass structures ?

But happily these monstrous gardens are but few —I only know of or have seen two, but I hope never to see another.

Nothing is more satisfactory than to see the well-designed and well-organised garden of the large country house, whose master loves his garden, and has good taste and a reasonable amount of leisure.

I think that the first thing in such a place is to have large unbroken lawn spaces—all the better if they are continuous, passing round the south and west sides of the house. I am supposing a house of the best class, but not necessarily of the largest size. Immediately adjoining the house, except for the few feet needed for a border for climbing plants, is a broad walk, dry and smooth, and perfectly level from end to end. This, in the case of many houses, and nearly always with good effect, is raised two or three feet above the garden ground, and if the architecture of the house demands it has a retaining wall surrounded by a balustrade of

M

masonry and wrought stone. Broad and shallow stone
steps lead down to the turf both at the end of the
walk and in the middle of the front of the house, the
wider and shallower the better, and at the foot of the
wall may be a narrow border for a few climbing plants
that will here and there rise above the coping of the
parapet. I do not think it desirable where there are
stone balusters or other distinct architectural features
to let them be smothered with climbing plants, but that
there should be, say, a *Pyrus japonica* or an Escallonia,
and perhaps a white Jasmine, and on a larger space
perhaps a cut-leaved or a Claret Vine. Some of the
best effects of the kind I have seen were where the
bush, being well established, rose straight out of the
grass, the border being unnecessary except just at the
beginning.

The large lawn space I am supposing stretches away
a good distance from the house, and is bounded on the
south and west by fine trees; away beyond that is all
wild wood. On summer afternoons the greater part
of the lawn expanse is in cool shade, while winter
sunsets show through the tree stems. Towards the
south-east the wood would pass into shrub plantations,
and farther still into garden and wild orchard (of
which I shall have something to say presently). At
this end of the lawn would be the brilliant parterre
of bedded plants, seen both from the shaded lawn and
from the terrace, which at this end forms part of its
design. Beyond the parterre would be a distinct

GARLAND-ROSE WREATHING THE END OF A TERRACE WALL.

division from the farther garden, either of Yew or
Box hedge, with bays for seats, or in the case of a
change of level, of another terrace wall. The next
space beyond would be the main garden for hardy
plants, at its southern end leading into the wild
orchard. This would be the place for the free garden
or the reserve garden, or for any of the many delightful ways in which hardy flowers can be used; and if
it happened by good fortune to have a stream or any
means of having running water, the possibilities of
beautiful gardening would be endless.

Beyond this again would come the kitchen garden,
and after that the stables and the home farm. If the
kitchen garden had a high wall, and might be entered
on this side by handsome wrought-iron gates, I would
approach it from the parterre by a broad grass walk
bounded by large Bay trees at equal intervals to right
and left. Through these to the right would be seen
the free garden of hardy flowers.

For the kitchen garden a space of two acres would
serve a large country house with all that is usually
grown within walls, but there should always be a good
space outside for the rougher vegetables, as well as a
roomy yard for compost, pits and frames, and rubbish.

And here I wish to plead on behalf of the gardener
that he should have all reasonable comforts and con-
veniences. Nothing is more frequent, even in good
places, than to find the potting and tool sheds screwed
away into some awkward corner, badly lighted, much

too small, and altogether inadequate, and the pits and frames scattered about and difficult to get at. Nothing is more wasteful of time, labour, or temper. The working parts of a large garden form a complicated organisation, and if the parts of the mechanism do not fit and work well, and are not properly eased and oiled, still more, if any are missing, there must be disastrous friction and damage and loss of power. In designing garden buildings, I always strongly urge in connection with the heating system a warmed potting shed and a comfortable messroom for the men, and over this a perfectly dry loft for drying and storing such matters as shading material, nets, mats, ropes, and sacks. If this can be warmed, so much the better. There must also be a convenient and quite frost-proof place for winter storing of vegetable roots and such plants as Dahlias, Cannas, and Gladiolus; and also a well-lighted and warmed workshop for all the innumerable jobs put aside for wet weather, of which the chief will be re-painting and glazing of lights, repairing implements, and grinding and setting tools. This shop should have a carpenter's bench and screw, and a smith's anvil, and a proper assortment of tools. Such arrangements, well planned and thought out, will save much time and loss of produce, besides helping to make all the people employed more comfortable and happy.

I think that a garden should never be large enough to be tiring, that if a large space has to be dealt

with, a great part had better be laid out in wood.
Woodland is always charming and restful and en-
duringly beautiful, and then there is an intermediate
kind of woodland that should be made more of—woodland of the orchard type. Why is the orchard put
out of the way, as it generally is, in some remote
region beyond the kitchen garden and stables? I
should like the lawn, or the hardy flower garden, or
both, to pass directly into it on one side, and to plant
a space of several acres, not necessarily in the usual
way, with orchard standards twenty-five feet apart in
straight rows (though in many places the straight rows
might be best), but to have groups and even groves
of such things as Medlars and Quinces, Siberian and
Chinese Crabs, Damsons, Prunes, Service trees, and
Mountain Ash, besides Apples, Pears, and Cherries,
in both standard and bush forms. Then alleys of
Filbert and Cob-nut, and in the opener spaces tangles
or brakes of the many beautiful bushy things allied to
the Apple and Plum tribe—*Cydonia* and *Prunus triloba*
and *Cratœgus* of many kinds (some of them are tall
bushes or small trees with beautiful fruits); and the
wild Blackthorn, which, though a plum, is so nearly
related to pear that pears may be grafted on it. And
then brakes of Blackberries, especially of the Parsley-
leaved kind, so free of growth and so generous of
fruit. How is it that this fine native plant is almost
invariably sold in nurseries as an American bramble?
If I am mistaken in this I should be glad to be

corrected, but I believe it to be only the cut-leaved variety of the native *Rubus affinis*.

I have tried the best of the American kinds, and with the exception of one year, when I had a few fine fruits from Kittatinny, they have been a failure, whereas invariably when people have told me that their American Blackberries have fruited well, I have found them to be the Parsley-leaved.

Some members of the large Rose-Apple-Plum tribe grow to be large forest trees, and in my wild orchard they would go in the farther parts. The Bird-cherry (*Prunus padus*) grows into a tree of the largest size. A Mountain Ash will sometimes have a trunk two feet in diameter, and a head of a size to suit. The American kind, its near relation, but with larger leaves and still grander masses of berries, is a noble small tree ; and the native white Beam should not be forgotten, and choice places should be given to Amelanchier and the lovely double Japan Apple (*Pyrus malus floribunda*). To give due space and effect to all these good things my orchard garden would run into a good many acres, but every year it would be growing into beauty and profit. The grass should be left rough, and plentifully planted with Daffodils, and with Cowslips if the soil is strong. The grass would be mown and made into hay in June, and perhaps mown once more towards the end of September. Under the nut-trees would be Primroses and the garden kinds of wood Hyacinths and Dogtooth Violets and Lily of the Valley, and perhaps Snowdrops, or any of the smaller bulbs that most commended themselves to the taste of the master.

Such an orchard garden, well-composed and beautifully grouped, always with that indispensable quality of good " drawing," would not only be a source of unending pleasure to those who lived in the place, but a valuable lesson to all who saw it; for it would show the value of the simple and sensible ways of using a certain class of related trees and bushes, and of using them with a deliberate intention of making the best of them, instead of the usual meaningless-nohow way of planting. This, in nine cases out of ten, means either ignorance or carelessness, the planter not caring enough about the matter to take the trouble to find out what is best to be done, and being quite satisfied with a mixed lot of shrubs, as offered in nursery sales, or with the choice of the nurseryman. I do not presume to condemn all mixed planting, only stupid and ignorant mixed planting. It is not given to all people to take their pleasures alike; and I have in my mind four gardens, all of the highest interest, in which the planting is all mixed; but then the mixture is of admirable ingredients, collected and placed on account of individual merit, and a ramble round any one of these in company with its owner is a pleasure and a privilege that one cannot prize too highly. Where the garden is of such large extent that experimental planting is made with a good number of one good thing

at a time, even though there was no premeditated
intention of planting for beautiful effect, the fact of
there being enough plants to fall into large groups,
and to cover some extent of ground, produces numbers
of excellent results. I remember being struck with
this on several occasions when I have had the happiness of visiting Mr. G. F. Wilson's garden at Wisley, a
garden which I take to be about the most instructive
it is possible to see. In one part, where the foot of
the hill joined the copse, there were hosts of lovely
things planted on a succession of rather narrow banks.
Almost unthinkingly I expressed the regret I felt that
so much individual beauty should be there without
an attempt to arrange it for good effect. Mr. Wilson
stopped, and looking at me straight with a kindly
smile, said very quietly, "That is your business, not
mine." In spite of its being a garden whose first
object is trial and experiment, it has left in my
memory two pictures, among several lesser ones, of
plant-beauty that will stay with me as long as I can
remember anything, one an autumn and one a spring
picture—the hedge of *Rosa rugosa* in full fruit, and a
plantation of *Primula denticulata*. The Primrose was
on a bit of level ground, just at the outer and inner
edges of the hazel copse. The plants were both
grouped and thinly sprinkled, just as nature plants—
possibly they grew directly there from seed. They
were in superb and luxuriant beauty in the black
peaty-looking half-boggy earth, the handsome leaves

A Roadside Cottage Garden.

of the brilliant colour and large size that told of perfect health and vigour, and the large round heads of
pure lilac flower carried on strong stalks that must
have been fifteen inches high. I never saw it so
happy and so beautiful. It is a plant I much admire,
and I do the best I can for it on my dry hill; but
the conditions of my garden do not allow of any
approach to the success of the Wisley plants; still I
have treasured that lesson among many others I have
brought away from that good garden, and never fail to
advise some such treatment when I see the likely home
for it in other places.

Some of the most delightful of all gardens are the
little strips in front of roadside cottages. They have
a simple and tender charm that one may look for in
vain in gardens of greater pretension. And the old
garden flowers seem to know that there they are seen
at their best; for where else can one see such Wallflowers, or Double Daisies, or White Rose bushes;
such clustering masses of perennial Peas, or such well-
kept flowery edgings of Pink, or Thrift, or London
Pride ?

Among a good many calls for advice about laying
out gardens, I remember an early one that was of
special interest. It was the window-box of a factory
lad in one of the great northern manufacturing towns.
He had advertised in a mechanical paper that he
wanted a tiny garden, as full of interest as might be,
in a window-box; he knew nothing—would somebody

help him with advice? So advice was sent and the
box prepared. If I remember rightly the size was
three feet by ten inches. A little later the post
brought him little plants of mossy and silvery saxifrages, and a few small bulbs. Even some stones were
sent, for it was to be a rock-garden, and there were to
be two hills of different heights with rocky tops, and a
longish valley with a sunny and a shady side.

It was delightful to have the boy's letters, full of
keen interest and eager questions, and only difficult to
restrain him from killing his plants with kindness, in
the way of liberal doses of artificial manure. The
very smallness of the tiny garden made each of its
small features the more precious. I could picture
his feeling of delightful anticipation when he saw the
first little bluish blade of the Snowdrop patch pierce
its mossy carpet. Would it, could it really grow into
a real Snowdrop, with the modest, milk-white flower
and the pretty green hearts on the outside of the inner
petals, and the clear green stripes within? and would it
really nod him a glad good-morning when he opened
his window to greet it? And those few blunt reddish
horny-looking snouts just coming through the ground,
would they really grow into the brilliant blue of the
early Squill, that would be like a bit of midsummer
sky among the grimy surroundings of the attic window,
and under that grey, soot-laden northern sky? I
thought with pleasure how he would watch them in
spare minutes of the dinner-hour spent at home, and

think of them as he went forward and back to his
work, and how the remembrance of the tender beauty
of the full-blown flower would make him glad, and lift
up his heart while "minding his mule" in the busy
restless mill.

CHAPTER XV

BEGINNING AND LEARNING

The ignorant questioner—Beginning at the end—An example—
Personal experience—Absence of outer help—Johns' "Flowers
of the Field"—Collecting plants—Nurseries near London—
Wheel-spokes as labels—Garden friends—Mr. Robinson's
"English Flower-Garden"—Mr. Nicholson's "Dictionary of
Gardening"—One main idea desirable—Pictorial treatment—
Training in fine art—Adapting from Nature—Study of colour—
Ignorant use of the word "artistic."

MANY people who love flowers and wish to do some
practical gardening are at their wit's end to know what
to do and how to begin. Like a person who is on
skates for the first time, they feel that, what with the
bright steel runners, and the slippery surface, and the
sense of helplessness, there are more ways of tumbling
about than of progressing safely in any one direction.
And in gardening the beginner must feel this kind of
perplexity and helplessness, and indeed there is a great
deal to learn, only it is pleasant instead of perilous,
and the many tumbles by the way only teach and do
not hurt. The first few steps are perhaps the most
difficult, and it is only when we know something of
the subject and an eager beginner comes with questions

that one sees how very many are the things that want
knowing.   And the more ignorant the questioner, the
more difficult it is to answer helpfully.   When one
knows, one cannot help presupposing some sort of
knowledge on the part of the querist, and where this
is absent the answer we can give is of no use.   The
ignorance, when fairly complete, is of such a nature
that the questioner does not know what to ask, and
the question, even if it can be answered, falls upon
barren ground.   I think in such cases it is better to
try and teach one simple thing at a time, and not to
attempt to answer a number of useless questions.   It
is disheartening when one has tried to give a careful
answer to have it received with an Oh! of boredom or
disappointment, as much as to say, You can't expect
me to take all that trouble; and there is the still more
unsatisfactory sort of applicant, who plies a string of
questions and will not wait for the answers!   The real
way is to try and learn a little from everybody and
from every place.   There is no royal road.   It is no
use asking me or any one else how to dig—I mean
sitting indoors and asking it.   Better go and watch a
man digging, and then take a spade and try to do it,
and go on trying till it comes, and you gain the knack
that is to be learnt with all tools, of doubling the power
and halving the effort; and meanwhile you will be
learning other things, about your own arms and legs
and back, and perhaps a little robin will come and
give you moral support, and at the same time keep a

sharp look-out for any worms you may happen to turn up; and you will find out that there are all sorts of ways of learning, not only from people and books, but from sheer trying.

I remember years ago having to learn to use the blow-pipe, for soldering and other purposes connected with work in gold and silver. The difficult part of it is to keep up the stream of air through the pipe while you are breathing the air in; it is easy enough when you only want a short blast of a few seconds, within the compass of one breath or one filling of the bellows (lungs), but often one has to go on blowing through several inspirations. It is a trick of muscular action. My master who taught me never could do it himself, but by much trying one day I caught the trick.

The grand way to learn, in gardening as in all things else, is to wish to learn, and to be determined to find out—not to think that any one person can wave a wand and give the power and knowledge. And there will be plenty of mistakes, and there must be, just as children must pass through the usual childish complaints. And some people make the mistake of trying to begin at the end, and of using recklessly what may want the utmost caution, such, for instance, as strong chemical mixtures.

Some ladies asked me why their plant had died. They had got it from the very best place, and they were sure they had done their very best for it, and— there it was, dead. I asked what it was, and how they had treated it. It was some ordinary border plant, whose identity I now forget; they had made a nice hole with their new trowel, and for its sole benefit they had bought a tin of Concentrated Fertiliser. This they had emptied into the hole, put in the plant, and covered it up and given it lots of water, and—it had died! And yet these were the best and kindest of women, who would never have dreamed of feeding a new-born infant on beefsteaks and raw brandy. But they learned their lesson well, and at once saw the sense when I pointed out that a plant with naked roots just taken out of the ground or a pot, removed from one feeding-place and not yet at home in another, or still more after a journey, with the roots only wrapped in a little damp moss and paper, had its feeding power suspended for a time, and was in the position of a helpless invalid. All that could be done for it then was a little bland nutriment of weak slops and careful nursing; if the planting took place in the summer it would want shading and only very gentle watering, until firm root-hold was secured and root-appetite became active, and that in rich and well-prepared garden ground such as theirs strong artificial manure was in any case superfluous.

When the earlier ignorances are overcome it becomes much easier to help and advise, because there is more common ground to stand on. In my own case, from quite a small child, I had always seen gardening going on, though not of a very interesting kind. Nothing much was thought of but bedding plants, and there was a rather large space on each side of the house for these, one on gravel and one on turf. But I had my own little garden in a nook beyond the shrubbery, with a seat shaded by a *Boursault elegans* Rose, which I thought then, and still think, one of the loveliest of its kind. But my first knowledge of hardy plants came through wild ones. Some one gave me that excellent book, the Rev. C. A. Johns' " Flowers of the Field." For many years I had no one to advise me (I was still quite small) how to use the book, or how to get to know (though it stared me in the face) how the plants were in large related families, and I had not the sense to do it for myself, nor to learn the introductory botanical part, which would have saved much trouble afterwards; but when I brought home my flowers I would take them one by one and just turn over the pages till I came to the picture that looked something like. But in this way I got a knowledge of individuals, and afterwards the idea of broad classification and relationship of genera to species may have come all the easier. I always think of that book as the most precious gift I ever received. I distinctly trace to its teaching my first firm steps in the path of plant knowledge, and the feeling of assured comfort I had afterwards in recognising the kinds when I came to collect garden plants; for at that time I had no other garden book, no means of access to botanic gardens or private collections, and no helpful adviser.

One copy of "Johns" I wore right out; I have now two, of which one is in its second binding, and is always near me for reference. I need hardly say that this was long before the days of the "English Flower-Garden," or its helpful predecessor, "Alpine Plants."

By this time I was steadily collecting hardy garden plants wherever I could find them, mostly from cottage gardens. Many of them were still unknown to me by name, but as the collection increased I began to compare and discriminate, and of various kinds of one plant to throw out the worse and retain the better, and to train myself to see what made a good garden plant, and about then began to grow the large yellow and white bunch Primroses, whose history is in another chapter. And then I learnt that there were such places (though then but few) as nurseries, where such plants as I had been collecting in the cottage gardens, and even better, were grown. And I went to Osborne's at Fulham (now all built over), and there saw the original tree of the fine Ilex known as the Fulham Oak, and several spring-flowering bulbs I had never seen before, and what I felt sure were numbers of desirable summer-flowering plants, but not then in bloom. Soon after this I began to learn something about Daffodils, and enjoyed much kind help from Mr. Barr, visiting his nursery (then at Tooting) several times, and sometimes combining a visit to Parker's nursery just over the way, a perfect paradise of good hardy plants. I shall never forget my first sight here

N

of the Cape Pondweed (*Aponogeton distachyon*) in full flower and great vigour in the dipping tanks, and over-flowing from them into the ditches.

Also I was delighted to see the use as labels of old wheel-spokes. I could not help feeling that if one had been a spoke of a cab-wheel, and had passed all one's working life in being whirled and clattered over London pavements, defiled with street mud, how pleasant a way to end one's days was this; to have one's felloe end pointed and dipped in nice wholesome rot-resisting gas-tar and thrust into the quiet cool earth, and one's nave end smoothed and painted and inscribed with some such soothing legend as *Vinca minor* or *Dianthus fragrans!*

Later I made acquaintance with several of the leading amateur and professional gardeners, and with Mr. Robinson, and to their good comradeship and kindly willingness to let me "pick their brains" I owe a great advance in garden lore. Moreover, what began by the drawing together of a common interest has grown into a still greater benefit, for several acquaintances so made have ripened into steady and much-valued friendships. It has been a great interest to me to have had the privilege of watching the gradual growth, through its several editions, of Mr. Robinson's "English Flower-Garden," the one best and most helpful book of all for those who want to know about hardy flowers, offering as it does in the clearest and easiest way a knowledge of the garden-treasures of the temperate world. No one who has not had occasional glimpses behind the scenes can know how much labour and thought such a book represents, to say nothing of research and practical experiment, and of the trouble and great expense of producing the large amount of pictorial illustration. Another book, though on quite different lines, that I find most useful is Mr. Nicholson's "Illustrated Dictionary of Gardening," in eight handy volumes. It covers much the same ground as the useful old Johnson's "Gardener's Dictionary," but is much more complete and comprehensive, and is copiously illustrated with excellent wood-cuts. It is the work of a careful and learned botanist, treating of all plants desirable for cultivation from all climates, and teaching all branches of practical horticulture and such useful matters as means of dealing with insect pests. The old "Johnson" is still a capital book in one volume; mine is rather out of date, being the edition of 1875, but it has been lately revised and improved. It would be delightful to possess, or to have easy access to, a good botanical library; still, for all the purposes of the average garden lover, these books will suffice.

I think it is desirable, when a certain degree of knowledge of plants and facility of dealing with them has been acquired, to get hold of a clear idea of what one most wishes to do. The scope of the subject is so wide, and there are so many ways to choose from, that having one general idea helps one to concentrate thought and effort that would otherwise be wasted by being diluted and dribbled through too many probable channels of waste.

Ever since it came to me to feel some little grasp of knowledge of means and methods, I have found that my greatest pleasure, both in garden and woodland, has been in the enjoyment of beauty of a pictorial kind. Whether the picture be large as of a whole landscape, or of lesser extent as in some fine single group or effect, or within the space of only a few inches as may be seen in some happily-disposed planting of Alpines, the intention is always the same; or whether it is the grouping of trees in the wood by the removal of those whose lines are not wanted in the picture, or in the laying out of broad grassy ways in woody places, or by ever so slight a turn or change of direction in a wood path, or in the alteration of some arrangement of related groups for form or for massing of light and shade, or for any of the many local conditions that guide one towards forming a decision, the intention is still always the same—to try and make a beautiful garden-picture. And little as I can as yet boast of being able to show anything like the number of these I could wish, yet during the flower-year there is generally something that at least in part answers to the effort.

I do not presume to urge the acceptance of my own particular form of pleasure in a garden on those to whom, from different temperament or manner of education, it would be unwelcome; I only speak of what I feel, and to a certain degree understand; but I had the advantage in earlier life of some amount of training in appreciation of the fine arts, and this, working upon an inborn feeling of reverent devotion to things of the highest beauty in the works of God, has helped me to an understanding of their divinely-inspired interpretations by the noblest minds of men, into those other forms that we know as works of fine art.

And so it comes about that those of us who feel and understand in this way do not exactly attempt to imitate Nature in our gardens, but try to become well acquainted with her moods and ways, and then discriminate in our borrowing, and so interpret her methods as best we may to the making of our garden-pictures.

I have always had great delight in the study of colour, as the word is understood by artists, which again is not a positive matter, but one of relation and proportion.   And when one hears the common chatter about "artistic colours," one receives an unpleasant impression about the education and good taste of the speaker; and one is reminded of an old saying which treats of the unwisdom of rushing in "where angels fear to tread," and of regret that a good word should be degraded by misuse.   It may be safely said that no colour can be called artistic in itself; for, in the first place, it is bad English, and in the second, it is nonsense.   Even if the first objection were waived, and the second condoned, it could only be used in a secondary sense, as signifying something that is useful and suitable and right in its place.   In this limited sense the scarlet of the soldier's coat, and of the pillar-box and mail-cart, and the bright colours of flags, or of the port and starboard lights of ships, might be said to be just so far "artistic" (again if grammar would allow), as they are right and good in their places.   But then those who use the word in the usual ignorant, random way have not even this simple conception of its meaning.   Those who know nothing about colour in the more refined sense (and like a knowledge of everything else it wants learning) get no farther than to enjoy it only when most crude and garish—when, as George Herbert says, it "bids the rash gazer wipe his eye," or when there is some violent opposition of complementary colour—forgetting, or not knowing, that though in detail the objects brought together may make each other appear brighter, yet in the mass, and especially when mixed up, the one actually neutralises the other.   And they have no idea of using the colour of flowers as precious jewels in a setting of quiet environment, or of suiting the colour of flowering groups to that of the neighbouring foliage, thereby enhancing the value of both, or of massing related or harmonious colourings so as to lead up to the most powerful and brilliant effects; and yet all these are just the ways of employing colour to the best advantage.

But the most frequent fault, whether in composition or in colour, is the attempt to crowd too much into the picture; the simpler effect obtained by means of temperate and wise restraint is always the more telling.

CHAPTER XVI

THE FLOWER-BORDER AND PERGOLA

The flower-border—The wall and its occupants—*Choisya ternata*—
Nandina—Canon Ellacombe's garden—Treatment of colour-
masses—Arrangement of plants in the border—Dahlias and
Cannas—Covering bare places—The pergola—How made—
Suitable climbers—Arbours of trained Planes—Garden houses.

I HAVE a rather large " mixed border of hardy flowers."
It is not quite so hopelessly mixed as one generally
sees, and the flowers are not all hardy ; but as it is a
thing everybody rightly expects, and as I have been
for a good many years trying to puzzle out its wants
and ways, I will try and describe my own and its surroundings.

There is a sandstone wall of pleasant colour at the
back, nearly eleven feet high.   This wall is an important feature in the garden, as it is the dividing line
between the pleasure garden and the working garden ;
also, it shelters the pleasure garden from the sweeping
blasts of wind from the north-west, to which my ground
is much exposed, as it is all on a gentle slope, going
downward towards the north.   At the foot of the wall
is a narrow border three feet six inches wide, and then
a narrow alley, not a made path, but just a way to go

A FLOWER-BORDER IN JUNE.

along for tending the wall shrubs, and for getting at the back of the border. This little alley does not show from the front. Then the main border, fourteen feet wide and two hundred feet long. About three-quarters of the way along a path cuts through the border, and passes by an arched gateway in the wall to the Pæony garden and the working garden beyond. Just here I thought it would be well to mound up the border a little, and plant with groups of Yuccas, so that at all times of the year there should be something to make a handsome full-stop to the sections of the border, and to glorify the doorway. The two extreme ends of the border are treated in the same way with Yuccas on rather lesser mounds, only leaving space beyond them for the entrance to the little alley at the back.

The wall and border face two points to the east of south, or, as a sailor would say, south-south-east, half-way between south and south-east. In front of the border runs a path seven feet wide, and where the border stops at the eastern end it still runs on another sixty feet, under the pergola, to the open end of a summer-house. The wall at its western end returns forward, square with its length, and hides out green-houses, sheds, and garden yard. The path in front of the border passes through an arch into this yard, but there is no view into the yard, as it is blocked by some Yews planted in a quarter-circle.

Though wall-space is always precious, I thought it

better to block out this shorter piece of return wall on
the garden side with a hedge of Yews. They are now
nearly the height of the wall, and will be allowed to
grow a little higher, and will eventually be cut into an
arch over the arch in the wall. I wanted the sombre
duskiness of the Yews as a rich, quiet background for
the brightness of the flowers, though they are rather disappointing in May and June, when their young shoots
are of a bright and lively green. At the eastern end of
the border there is no return wall, but another planting of Yews equal to the depth of the border. Notched
into them is a stone seat about ten feet long; as they
grow they will be clipped so as to make an arching
hood over the seat.

The wall is covered with climbers, or with non-
climbing shrubs treated as wall-plants. They do not
all want the wall for warmth or protection, but are
there because I want them there; because, thinking
over what things would look best and give me the
greatest pleasure, these came among them. All the
same, the larger number of the plants on the wall do
want it, and would not do without it. At the western
end, the only part which is in shade for the greater
part of the day, is a *Garrya elliptica*. So many of my
garden friends like a quiet journey along the wall to
see what is there, that I propose to do the like by my
reader; so first for the wall, and then for the border.
Beyond the *Garrya*, in the extreme angle, is a *Clematis
montana*. When the *Garrya* is more grown there will

PATHWAY ACROSS THE SOUTH BORDER IN JULY.

OUTSIDE VIEW OF THE BRICK PERGOLA SHOWN AT PAGE 214, AFTER
SIX YEARS' GROWTH.

not be much room left for the Clematis, but then it
will have become bare below, and can ramble over the
wall on the north side, and, in any case, it is a plant
with a not very long lifetime, and will be nearly or
quite worn out before its root-space is reached or
wanted by its neighbours. Next on the wall is the
beautiful Rose Acacia (*Robinia hispida*). It is perfectly
hardy, but the wood is so brittle that it breaks off
short with the slightest weight of wind or snow or
rain. I never could understand why a hardy shrub
was created so brittle, or how it behaves in its native
place. I look in my "Nicholson," and see that it
comes from North America. Now, North America
is a large place, and there may be in it favoured
spots where there is no snow, and only the very
gentlest rain, and so well sheltered that the wind only
blows in faintest breaths ; and to judge by its behaviour
in our gardens, all these conditions are necessary for
its well-being. This troublesome quality of brittleness
no doubt accounts for its being so seldom seen in
gardens. I began to think it hopeless when, after
three plantings in the open, it was again wrecked, but
at last had the happy idea of training it on a wall.
Even there, though it is looked over and tied in twice
a year, a branch or two often gets broken. But I
do not regret having given it the space, as the wall
could hardly have had a better ornament, so beautiful
are its rosy flower-clusters and pale-green leaves. As
it inclines to be leggy below, I have trained a Crimson

Rambler Rose over the lower part, tying it in to any bare places in the *Robinia*.

Next along the wall is *Solanum crispum*, much to be recommended in our southern counties. It covers a good space of wall, and every year shoots up some feet above it; indeed it is such a lively grower that it has to endure a severe yearly pruning. Every season it is smothered with its pretty clusters of potato-shaped bloom of a good bluish-lilac colour. After these I wanted some solid-looking dark evergreens, so there is a Loquat, with its splendid foliage equalling that of *Magnolia grandiflora*, and then Black Laurustinus, Bay, and Japan Privet; and from among this dark-leaved company shoots up the tender green of a Banksian Rose, grown from seed of the single kind, the gift of my kind friend Commendatore Hanbury, whose world-famed garden of La Mortola, near Ventimiglia, probably contains the most remarkable collection of plants and shrubs that have ever been brought together by one man. This Rose has made good growth, and a first few flowers last year—seedling Roses are slow to bloom—lead me to expect a good show next season.

In the narrow border at the foot of the wall is a bush of *Raphiolepis ovata*, always to me an interesting shrub, with its thick, roundish, leathery leaves and white flower-clusters, also bushes of Rosemary, some just filling the border, and some trained up the wall. Our Tudor ancestors were fond of Rosemary-covered walls, and I have seen old bushes quite ten feet high on the garden walls of Italian monasteries. Among the Rosemaries I always like, if possible, to "tickle in" a China Rose or two, the tender pink of the Rose seems to go so well with the dark but dull-surfaced Rosemary. Then still in the wall-border comes a long straggling mass of that very pretty and interesting herbaceous Clematis, *C. Davidiana*. The colour of its flower always delights me; it is of an unusual kind of greyish-blue, of very tender and lovely quality. It does well in this warm border, growing about three feet high. Then on the wall come *Pyrus Maulei* and *Chimonanthus*, Claret-Vine, and the large-flowered *Ceanothus* Gloire de Versailles, hardy *Fuchsia*, and *Magnolia Soulangeana*, ending with a big bush of *Choisya ternata*, and rambling above it a very fine kind of *Bignonia grandiflora*.

Then comes the archway, flanked by thick buttresses. A Choisya was planted just beyond each of these, but it has grown wide and high, spreading across the face of the buttress on each side, and considerably invading the pathway. There is no better shrub here than this delightful Mexican plant; its long whippy roots ramble through our light soil with every sign of enjoyment; it always looks clean and healthy and well dressed, and as for its lovely and deliciously sweet flowers, we cut them by the bushel, and almost by the faggot, and the bushes scarcely look any emptier.

Beyond the archway comes the shorter length of wall and border. For convenience I planted all slightly tender things together on this bit of wall and

border; then we make one job of covering the whole
with fir-boughs for protection in winter.   On the wall
are *Piptanthus nepalensis, Cistus ladaniferus, Edwardsia
grandiflora,* and another Loquat, and in the border a number of Hydrangeas, *Clerodendron fœtidium, Crinums,* and
Nandina domestica, the Chinese so-called sacred Bamboo.
It is not a Bamboo at all, but allied to *Berberis;* the
Chinese plant it for good luck near their houses.   If
it is as lucky as it is pretty, it ought to do one good !
I first made acquaintance with this beautiful plant in
Canon Ellacombe's most interesting garden at Bitton, in
Gloucestershire, where it struck me as one of the most
beautiful growing things I had ever seen, the beauty
being mostly in the form and colouring of the leaves.
It is not perhaps a plant for everybody, and barely
hardy; it seems slow to get hold, and its full beauty
only shows when it is well established, and throws up its
wonderfully-coloured leaves on tall bamboo-like stalks.

There is nothing much more difficult to do in outdoor gardening than to plant a mixed border well, and
to keep it in beauty throughout the summer.   Every
year, as I gain more experience, and, I hope, more
power of critical judgment, I find myself tending
towards broader and simpler effects, both of grouping
and colour.   I do not know whether it is by individual
preference, or in obedience to some colour-law that I can
instinctively feel but cannot pretend even to understand,
and much less to explain, but in practice I always find
more satisfaction and facility in treating the warm

colours (reds and yellows) in graduated harmonies, culminating into gorgeousness, and the cool ones in contrasts; especially in the case of blue, which I like to use either in distinct but not garish contrasts, as of full blue with pale yellow, or in separate cloud-like harmonies, as of lilac and pale purple with grey foliage. I am never so much inclined to treat the blues, purples, and lilacs in gradations together as I am the reds and yellows. Purples and lilacs I can put together, but not these with blues; and the pure blues always seem to demand peculiar and very careful treatment.

The western end of the flower-border begins with the low bank of Yuccas, then there are some rather large masses of important grey and glaucous foliage and pale and full pink flower. The foliage is mostly of the Globe Artichoke, and nearer the front of *Artemisia* and *Cineraria maritima*. Among this, pink Canterbury Bell, Hollyhock, Phlox, Gladiolus, and Japan Anemone, all in pink colourings, will follow one another in due succession. Then come some groups of plants bearing whitish and very pale flowers, *Polygonum compactum*, *Aconitum lycoctonum*, Double Meadowsweet, and other Spiræas, and then the colour passes to pale yellow of Mulleins, and with them the palest blue Delphiniums. Towards the front is a wide planting of *Iris pallida dalmatica*, its handsome bluish foliage showing as outstanding and yet related masses with regard to the first large group of pale foliage. Then comes the pale-yellow *Iris flavescens*, and meanwhile the group of Delphinium deepens into those of a fuller blue colour, though none of the darkest are here. Then more pale yellow of Mullein, Thalictrum, and Paris Daisy, and so the colour passes to stronger yellows. These change into orange, and from that to brightest scarlet and crimson, coming to the fullest strength in the Oriental Poppies of the earlier year, and later in Lychnis, Gladiolus, Scarlet Dahlia, and Tritoma. The colour-scheme then passes again through orange and yellow to the paler yellows, and so again to blue and warm white, where it meets one of the clumps of Yuccas flanking the path that divides this longer part of the border from the much shorter piece beyond. This simple procession of colour arrangement has occupied a space of a hundred and sixty feet, and the border is all the better for it.

The short length of border beyond the gateway has again Yuccas and important pale foliage, and a preponderance of pink bloom, Hydrangea for the most part; but there are a few tall Mulleins, whose pale-yellow flowers group well with the ivory of the Yucca spikes and the clear pink of the tall Hollyhocks. These all show up well over the masses of grey and glaucous foliage, and against the rich darkness of dusky Yew.

Dahlias and Cannas have their places in the mixed border. When it is being dismantled in the late autumn all bare places are well dug and enriched, so that when it comes to filling-up time, at the end of May, I know that every spare bit of space is ready, and at the time of preparation I mark places for special Dahlias, according to colour, and for groups of the tall Cannas where I want grand foliage.

There are certain classes of plants that are quite indispensable, but that leave a bare or shabby-looking place when their bloom is over. How to cover these places is one of the problems that have to be solved. The worst offender is Oriental Poppy; it becomes unsightly soon after blooming, and is quite gone by midsummer. I therefore plant *Gypsophila paniculata* between and behind the Poppy groups, and by July there is a delicate cloud of bloom instead of large bare patches. *Eryngium Olivieranum* has turned brown by the beginning of July, but around the group some Dahlias have been planted, that will be gradually trained down over the space of the departed Sea-Holly, and other Dahlias are used in the same way to mask various weak places.

There is a perennial Sunflower, with tall black stems, and pale-yellow flowers quite at the top, an old garden sort, but not very good as usually grown; this I find of great value to train down, when it throws up a short flowering stem from each joint, and becomes a spreading sheet of bloom.

One would rather not have to resort to these artifices of sticking and training; but if a certain effect is wanted, all such means are lawful, provided that nothing looks stiff or strained or unsightly; and it is pleasant to exercise ingenuity and to invent ways to

o

meet the needs of any case that may arise. But like everything else, in good gardening it must be done just right, and the artist-gardener finds that hardly the placing of a single plant can be deputed to any other hand than his own; for though, when it is done, it looks quite simple and easy, he must paint his own picture himself—no one can paint it for him.

I have no dogmatic views about having in the so-called hardy flower-border none but hardy flowers. All flowers are welcome that are right in colour, and that make a brave show where a brave show is wanted. It is of more importance that the border should be handsome than that all its occupants should be hardy. Therefore I prepare a certain useful lot of half-hardy annuals, and a few of what have come to be called bedding-plants. I like to vary them a little from year to year, because in no one season can I get in all the good flowers that I should like to grow; and I think it better to leave out some one year and have them the next, than to crowd any up, or to find I have plants to put out and no space to put them in. But I nearly always grow these half-hardy annuals; orange African Marigold, French Marigold, sulphur Sunflower, orange and scarlet tall Zinnia, Nasturtiums, both dwarf and trailing, *Nicotiana affinis*, Maize, and Salpi-glossis. Then Stocks and China Asters. The Stocks are always the large white and flesh-coloured summer kinds, and the Asters, the White Comet, and one of the blood-red or so-called scarlet sorts.

END OF FLOWER-BORDER AND ENTRANCE OF PERGOLA.

SOUTH BORDER DOOR AND YUCCAS IN AUGUST.

Then I have yellow Paris Daisies, *Salvia patens*, Heliotrope, *Calceolaria amplexicaulis*, Geraniums, scarlet and salmon-coloured and ivy-leaved kinds, the best of these being the pink Madame Crousse.

The front edges of the border are also treated in rather a large way. At the shadier end there is first a long straggling bordering patch of *Anemone sylvestris*. When it is once above ground the foliage remains good till autumn, while its soft white flower comes right with the colour of the flowers behind. Then comes a long and large patch of the larger kind of *Megasea cordifolia*, several yards in length, and running back here and there among taller plants. I am never tired of admiring the fine solid foliage of this family of plants, remaining, as it does, in beauty both winter and summer, and taking on a splendid winter colouring of warm red bronze. It is true that the flowers of the two best-known kinds, *M. cordifolia* and *M. crassifolia*, have coarse-looking flowers of a strong and rank quality of pink colour, but the persistent beauty of the leaves more than compensates; and in the rather tenderer kind, *M. ligulata* and its varieties, the colour of the flower is delightful, of a delicate good pink, with almost scarlet stalks. There is nothing flimsy or temporary-looking about the Megaseas, but rather a sort of grave and monumental look that specially fits them for association with masonry, or for any place where a solid-looking edging or full-stop is wanted. To go back to those in the edge of the border: if the edging

threatens to look too dark and hard, I plant among
or just behind the plants that compose it pink or
scarlet Ivy Geranium or trailing Nasturtium, according to the colour demanded by the neighbouring group.
Heuchera Richardsoni is another good front-edge plant;
and when we come to the blue and pale-yellow group
there is a planting of *Funkia grandiflora*, whose fresh-
looking pale green leaves are delightful with the
brilliant light yellow of *Calceolaria amplexicaulis*, and
the farther-back planting of pale-blue Delphinium,
Mullein, and sulphur Sunflower; while the same colour
of foliage is repeated in the fresh green of the Indian
Corn. Small spaces occur here and there along the
extreme front edge, and here are planted little jewels
of colour, of blue Lobelia, or dwarf Nasturtium, or
anything of the colour that the place demands.

The whole thing sounds much more elaborate than
it really is; the trained eye sees what is wanted, and
the trained hand does it, both by an acquired instinct.
It is painting a picture with living plants.

I much enjoy the pergola at the end of the sunny
path. It is pleasant while walking in full sunshine,
and when that sunny place feels just a little too hot,
to look into its cool depth, and to feel that one has
only to go a few steps farther to be in shade, and to
feel that little air of wind that the moving summer
clouds say is not far off, and is only unfelt just here
because it is stopped by the wall. It feels wonderfully
dark at first, this gallery of cool greenery, passing into

it with one's eyes full of light and colour, and the open-sided summer-house at the end looks like a black cavern; but on going into it, and sitting down on one of its broad, low benches, one finds that it is a pleasant subdued light, just right to read by.

The pergola has two openings out of it on the right, and one on the left. The first way out on the right is straight into the nut-walk, which leads up to very near the house. The second goes up two or three low, broad steps made of natural sandstone flags, between groups of Ferns, into the Michaelmas Daisy garden. The opening on the left leads into a quiet space of grass the width of the flower and wall border (twenty feet), having only some peat-beds planted with Kalmia. This is backed by a Yew hedge in continuation of the main wall, and it will soon grow into a cool, quiet bit of garden, seeming to belong to the pergola. Now, standing midway in the length of the covered walk, with the eye rested and refreshed by the leafy half-light, on turning round again towards the border it shows as a brilliant picture through the bowery framing, and the value of the simple method of using the colours is seen to full advantage.

I do not like a mean pergola, made of stuff as thin as hop-poles. If means or materials do not admit of having anything better, it is far better to use these in some other simple way, of which there are many to choose from—such as uprights at even intervals, braced together with a continuous rail at about four feet from the ground, and another rail just clear of the ground, and some simple trellis of the smaller stuff between these two rails. This is always pretty at the back of a flower-border in any modest garden. But a pergola should be more seriously treated, and the piers at any rate should be of something rather large—either oak stems ten inches thick, or, better still, of fourteen-inch brickwork painted with lime-wash to a quiet stone-colour. In Italy the piers are often of rubble masonry, either round or square in section, coated with very coarse plaster, and lime-washed white. For a pergola of moderate size the piers should stand in pairs across the path, with eight feet clear between. Ten feet from pier to pier along the path is a good proportion, or anything from eight to ten feet, and they should stand seven feet two inches out of the ground. Each pair should be tied across the top with a strong beam of oak, either of the natural shape, or roughly adzed on the four faces; but in any case, the ends of the beams, where they rest on the top of the piers, should be adzed flat to give them a firm seat. If the beams are slightly curved or cambered, as most trunks of oak are, so much the better, but they must always be placed camber side up. The pieces that run along the top, with the length of the path, may be of any branching tops of oak, or of larch poles. These can easily be replaced as they decay; but the replacing of a beam is a more difficult matter, so that it is well to let them be fairly durable from the beginning.

STONE-BUILT PERGOLA WITH WROUGHT OAK BEAMS.

PERGOLA WITH BRICK PIERS AND BEAMS OF ROUGH OAK.

The climbers I find best for covering the pergola are Vines, Jasmine, Aristolochia, Virginia Creeper, and Wistaria. Roses are about the worst, for they soon run up leggy, and only flower at the top out of sight.

A sensible arrangement, allied to the pergola, and frequent in Germany and Switzerland, is made by planting young Planes, pollarding them at about eight feet from the ground, and training down the young growths horizontally till they have covered the desired roof-space.

There is much to be done in our better-class gardens in the way of pretty small structures thoroughly well-designed and built. Many a large lawn used every afternoon in summer as a family playground and place to receive visitors would have its comfort and usefulness greatly increased by a pretty garden-house, instead of the usual hot and ugly, crampy and uncomfortable tent. But it should be thoroughly well designed to suit the house and garden. A pigeon-cote would come well in the upper part, and the face or faces open to the lawn might be closed in winter with movable shutters, when it would make a useful store-place for garden seats and much else.

CHAPTER XVII

THE PRIMROSE GARDEN

IT must be some five-and-twenty years ago that I began to work at what I may now call my own strain of Primroses, improving it a little every year by careful selection of the best for seed. The parents of the strain were a named kind, called Golden Plover, and a white one, without name, that I found in a cottage garden. I had also a dozen plants about eight or nine years ago from a strong strain of Mr. Anthony Waterer's that was running on nearly the same lines; but a year later, when I had flowered them side by side, I liked my own one rather the best, and Mr. Waterer, seeing them soon after, approved of them so much that he took some to work with his own. I hold Mr. Waterer's strain in great admiration, and, though I tried for a good many years, never could come near him in red colourings. But as my own taste favoured the delicately-shaded flowers, and the ones most liked in the nursery seemed to be those with strongly contrasting eye, it is likely that the two strains may be working still farther apart.

They are, broadly speaking, white and yellow

EVENING IN THE PRIMROSE GARDEN.

varieties of the strong bunch-flowered or Polyanthus
kind, but they vary in detail so much, in form, colour,
habit, arrangement, and size of eye and shape of edge,
that one year thinking it might be useful to classify
them I tried to do so, but gave it up after writing out
the characters of sixty classes! Their possible variation seems endless. Every year among the seedlings
there appear a number of charming flowers with some
new development of size, or colour of flower, or beauty
of foliage, and yet all within the narrow bounds of—
white and yellow Primroses.

Their time of flowering is much later than that of
the true or single-stalked Primrose. They come into
bloom early in April, though a certain number of
poorly-developed flowers generally come much earlier,
and they are at their best in the last two weeks of
April and the first days of May. When the bloom
wanes, and is nearly overtopped by the leaves, the
time has come that I find best for dividing and replanting. The plants then seem willing to divide,
some almost falling apart in one's hands, and the new
roots may be seen just beginning to form at the base
of the crown. The plants are at the same time
relieved of the crowded mass of flower-stem, and,
therefore, of the exhausting effort of forming seed, a
severe drain on their strength. A certain number will
not have made more than one strong crown, and a few
single-crown plants have not flowered; these, of course,
do not divide. During the flowering time I keep a

good look-out for those that I judge to be the most beautiful and desirable, and mark them for seed. These are also taken up, but are kept apart, the flower stems reduced to one or two of the most promising, and they are then planted in a separate place—some cool nursery corner. I find that the lifting and re-planting in no way checks the growth or well-being of the seed-pods.

I remember some years ago a warm discussion in the gardening papers about the right time to sow the seed. Some gardeners of high standing were strongly for sowing it as soon as ripe, while others equally trustworthy advised holding it over till March. I have tried both ways, and have satisfied myself that it is a matter for experiment and decision in individual gardens. As nearly as I can make out, it is well in heavy soils to sow when ripe, and in light ones to wait till March. In some heavy soils Primroses stand well for two years without division; whereas in light ones, such as mine, they take up the food within reach in a much shorter time, so that by the second year the plant has become a crowded mass of weak crowns that only throw up poor flowers, and are by then so much exhausted that they are not worth dividing afterwards. In my own case, having tried both ways, I find the March sown ones much the best.

The seed is sown in boxes in cold frames, and pricked out again into boxes when large enough to handle. The seedlings are planted out in June, when they seem to go on without any check whatever, and are just right for blooming next spring.

The Primrose garden is in a place by itself—a clearing half shaded by Oak, Chestnut, and Hazel. I always think of the Hazel as a kind nurse to Primroses; in the copses they generally grow together, and the finest Primrose plants are often nestled close in to the base of the nut-stool. Three paths run through the Primrose garden, mere narrow tracks between the beds, converging at both ends, something like the lines of longitude on a globe, the ground widening in the middle where there are two good-sized Oaks, and coming to a blunt point at each end, the only other planting near it being two other long-shaped strips of Lily of the Valley.

Every year, before replanting, the Primrose ground is dug over and well manured. All day for two days I sit on a low stool dividing the plants; a certain degree of facility and expertness has come of long practice. The "rubber" for frequent knife-sharpening is in a pail of water by my side; the lusciously fragrant heap of refuse leaf and flower-stem and old stocky root rises in front of me, changing its shape from a heap to a ridge, as when it comes to a certain height and bulk I back and back away from it. A boy feeds me with armfuls of newly-dug-up plants, two men are digging-in the cooling cow-dung at the farther end, and another carries away the divided plants tray by tray, and carefully replants them. The still air, with only the very gentlest south-westerly breath in it, brings up the mighty boom of the great ship guns from the old seaport, thirty miles away, and the pheasants answer to the sound as they do to thunder. The early summer air is of a perfect temperature, the soft coo of the wood-dove comes down from the near wood, the nightingale sings almost overhead, but—either human happiness may never be quite complete, or else one is not philosophic enough to contemn life's lesser evils, for—oh, the midges!

CHAPTER XVIII

COLOURS OF FLOWERS

I AM always surprised at the vague, not to say reckless, fashion in which garden folk set to work to describe the colours of flowers, and at the way in which quite wrong colours are attributed to them. It is done in perfect good faith, and without the least consciousness of describing wrongly. In many cases it appears to be because the names of certain substances have been used conventionally or poetically to convey the idea of certain colours. And some of these errors are so old that they have acquired a kind of respectability, and are in a way accepted without challenge. When they are used about familiar flowers it does not occur to one to detect them, because one knows the flower and its true colour; but when the same old error is used in the description of a new flower, it is distinctly misleading. For instance, when we hear of golden buttercups, we know that it means bright-yellow buttercups; but in the case of a new flower, or one not generally known, surely it is better and more accurate to say bright yellow at once. Nothing is more frequent in plant catalogues than " bright golden yellow," when bright yellow is meant. Gold is not bright yellow. I find that a gold piece laid on a gravel path, or against a sandy bank, nearly matches it in colour; and I cannot think of any flower that matches or even approaches the true colour of gold, though something near it may be seen in the pollen-covered anthers of many flowers. A match for gold may more nearly be found among dying beech leaves, and some dark colours of straw or dry grass bents, but none of these when they match the gold are bright yellow. In literature it is quite another matter; when the poet or imaginative writer says, " a field of golden buttercups," or " a golden sunset," he is quite right, because he appeals to our artistic perception, and in such case only uses the word as an image of something that is rich and sumptuous and glowing.

The same irrelevance of comparison seems to run through all the colours. Flowers of a full, bright-blue colour are often described as of a " brilliant amethystine blue." Why amethystine? The amethyst, as we generally see it, is a stone of a washy purple colour, and though there are amethysts of a fine purple, they are not so often seen as the paler ones, and I have never seen one even faintly approaching a really blue colour. What, therefore, is the sense of likening a flower, such as a Delphinium, which is really of a splendid pure-blue colour, to the duller and totally different colour of a third-rate gem?

Another example of the same slip-slop is the term

flame-coloured, and it is often preceded by the word
" gorgeous."   This contradictory mixture of terms is
generally used to mean bright scarlet.   When I look
at a flame, whether of fire or candle, I see that the
colour is a rather pale yellow, with a reddish tinge
about its upper forks, and side wings often of a bluish
white—no scarlet anywhere.   The nearest approach to
red is in the coals, not in the flame.   In the case of
the candle, the point of the wick is faintly red when
compared with the flame, but about the flame there is
no red whatever.   A distant bonfire looks red at night,
but I take it that the apparent redness is from seeing
the flames through damp atmosphere, just as the har-
vest-moon looks red when it rises.

And the strange thing is that in all these cases the
likeness to the unlike, and much less bright, colour is
given with an air of conferring the highest compliment
on the flower in question.   It is as if, wishing to praise
some flower of a beautiful blue, one called it a brilliant
slate-roof blue.   This sounds absurd, because it is
unfamiliar, but the unsuitability of the comparison is
scarcely greater than in the examples just quoted.

It seems most reasonable in describing the colour
of flowers to look out for substances whose normal
colour shows but little variation—such, for example, as
sulphur.   The colour of sulphur is nearly always the
same.   Citron, lemon, and canary are useful colour-
names, indicating different strengths of pure pale
yellow, inclining towards a tinge of the palest green.

Gentian-blue is a useful word, bringing to mind the piercingly powerful hue of the Gentianella. So also is turquoise-blue, for the stone has little variety of shade, and the colour is always of the same type. Forget-me-not blue is also a good word, meaning the colour of the native water Forget-me-not. Sky-blue is a little vague, though it has come by the " crystallising " force of usage to stand for a blue rather pale than full, and not far from that of the Forget-me-not; indeed, I seem to remember written passages in which the colours of flower and firmament were used reciprocally, the one in describing the other. Cobalt is a word sometimes used, but more often misused, for only water-colour painters know just what it represents, and it is of little use, as it so rarely occurs among flowers.

Crimson is a word to beware of; it covers such a wide extent of ground, and is used so carelessly in plant-catalogues, that one cannot know whether it stands for a rich blood colour or for a malignant magenta. For the latter class of colour the term amaranth, so generally used in French plant-lists, is extremely useful, both as a definition and a warning. Salmon is an excellent colour-word, copper is also useful, the two covering a limited range of beautiful colouring of the utmost value. Blood-red is also accurately descriptive. Terra-cotta is useful but indefinite, as it may mean anything between brick-red and buff. Red-lead, if it would be accepted as a

colour-word, would be useful, denoting the shades of
colour between the strongest orange and the palest
scarlet, frequent in the lightest of the Oriental Poppies.
Amber is a misleading word, for who is to know when it
means the transparent amber, whose colour approaches
that of resin, or the pale, almost opaque, dull-yellow
kind. And what is meant by coral-red? It is the
red of the old-fashioned dull-scarlet coral, or of the
pink kind more recently in favour.

The terms bronze and smoke may well be used in
their place, as in describing or attempting to describe
the wonderful colouring of such flowers as Spanish
Iris, and the varieties of Iris of the *squalens* section.
But often in describing a flower a reference to texture
much helps and strengthens the colour-word. I have
often described the modest little *Iris tuberosa* as a
flower made of green satin and black velvet. The
green portion is only slightly green, but is entirely
green satin, and the black of the velvet is barely black,
but is quite black-velvet-like. The texture of the
flower of *Ornithogalum nutans* is silver satin, neither
very silvery nor very satin-like, and yet so nearly
suggesting the texture of both that the words may
well be used in speaking of it. Indeed, texture plays
so important a part in the appearance of colour-sur-
face, that one can hardly think of colour without also
thinking of texture. A piece of black satin and a
piece of black velvet may be woven of the same batch
of material, but when the satin is finished and the

P

velvet cut, the appearance is often so dissimilar that they may look quite different in colour. A working painter is never happy if you give him an oil-colour pattern to match in distemper; he must have it of the same texture, or he will not undertake to get it like.

What a wonderful range of colouring there is in black alone to a trained colour-eye ! There is the dull brown-black of soot, and the velvety brown-black of the bean-flower's blotch; to my own eye, I have never found anything so entirely black in a natural product as the patch on the lower petals of *Iris iberica.* Is it not Ruskin who says of Velasquez, that there is more colour in his black than in many another painter's whole palette ? The blotch of the bean-flower appears black at first, till you look at it close in the sunlight, and then you see its rich velvety texture, so nearly like some of the brown-velvet markings on butterflies' wings. And the same kind of rich colour and texture occurs again on some of the tough flat half-round funguses, marked with shaded rings, that grow out of old posts, and that I always enjoy as lessons of lovely colour-harmony of grey and brown and black.

Much to be regretted is the disuse of the old word murrey, now only employed in heraldry. It stands for a dull red-purple, such as appears in the flower of the Virginian Allspice, and in the native Hound's-tongue, and often in seedling Auriculas. A fine strong-growing border Auricula was given to me by my valued friend the Curator of the Trinity College Botanic Garden,

Dublin, to which he had given the excellently descriptive name, " Old Murrey."

Sage-green is a good colour-word, for, winter or summer, the sage-leaves change but little.   Olive-green is not so clear, though it has come by use to stand for a brownish green, like the glass of a wine-bottle held up to the light, but perhaps bottle-green is the better word.   And it is not clear what part or condition of the olive is meant, for the ripe fruit is nearly black, and the tree in general, and the leaf in detail, are of a cool-grey colour.   Perhaps the colour-word is taken from the colour of the unripe fruit pickled in brine, as we see them on the table.   Grass-green any one may understand, but I am always puzzled by apple-green.   Apples are of so many different greens, to say nothing of red and yellow ; and as for pea-green, I have no idea what it means.

I notice in plant-lists the most reckless and indiscriminate use of the words purple, violet, mauve, lilac, and lavender, and as they are all related, I think they should be used with the greater caution.   I should say that mauve and lilac cover the same ground; the word mauve came into use within my recollection.   It is French for mallow, and the flower of the wild plant may stand as the type of what the word means.   Lavender stands for a colder or bluer range of pale purples, with an inclination to grey ; it is a useful word, because the whole colour of the flower spike varies so little.   Violet stands for the dark garden violet, and I always think of the grand colour of *Iris reticulata* as an example of a rich violet-purple. But purple equally stands for this, and for many shades redder.

Snow-white is very vague. There is nearly always so much blue about the colour of snow, from its crystalline surface and partial transparency, and the texture is so unlike that of any kind of flower, that the comparison is scarcely permissible. I take it that the use of "snow-white" is, like that of "golden-yellow," more symbolical than descriptive, meaning any white that gives an impression of purity. Nearly all white flowers are yellowish-white, and the comparatively few that are bluish-white, such, for example, as *Omphalodes verna*, are of a texture so different from snow that one cannot compare them at all. I should say that most white flowers are near the colour of chalk; for although the word chalky-white has been used in rather a contemptuous way, the colour is really a very beautiful warm white, but by no means an intense white. The flower that always looks to me the whitest is that of *Iberis sempervirens*. The white is dead and hard, like a piece of glazed stoneware, quite without play or variation, and hence uninteresting.

CHAPTER XIX

THE SCENTS OF THE GARDEN

THE sweet scents of a garden are by no means the least of its many delights. Even January brings *Chimonanthus fragrans*, one of the sweetest and strongest scented of the year's blooms—little half-transparent yellowish bells on an otherwise naked-looking wall shrub. They have no stalks, but if they are floated in a shallow dish of water, they last well for several days, and give off a powerful fragrance in a room.

During some of the warm days that nearly always come towards the end of February, if one knows where to look in some sunny, sheltered corner of a hazel copse, there will be sure to be some Primroses, and the first scent of the year's first Primrose is no small pleasure. The garden Primroses soon follow, and, meanwhile, in all open winter weather there have been Czar Violets and *Iris stylosa*, with its delicate scent, faintly violet-like, but with a dash of tulip. *Iris reticulata* is also sweet, with a still stronger perfume of the violet character. But of all Irises I know, the sweetest to smell is a later blooming one, *I. graminea*. Its small purple flowers are almost hidden among the

thick mass of grassy foliage which rises high above the
bloom; but they are worth looking for, for the sake
of the sweet and rather penetrating scent, which is
exactly like that of a perfectly-ripened plum.

All the scented flowers of the Primrose tribe are
delightful—Primrose, Polyanthus, Auricula, Cowslip.
The actual sweetness is most apparent in the Cowslip;
in the Auricula it has a pungency, and at the same
time a kind of veiled mystery, that accords with the
clouded and curiously-blended colourings of many of
the flowers.

Sweetbriar is one of the strongest of the year's
early scents, and closely following is the woodland
incense of the Larch, both freely given off and far-
wafted, as is also that of the hardy Daphnes. The
first quarter of the year also brings the bloom of most
of the deciduous Magnolias, all with a fragrance nearly
allied to that of the large one that blooms late in
summer, but not so strong and heavy.

The sweetness of a sun-baked bank of Wallflower
belongs to April. Daffodils, lovely as they are, must
be classed among flowers of rather rank smell, and yet
it is welcome, for it means spring-time, with its own
charm and its glad promise of the wealth of summer
bloom that is soon to come. The scent of the Jonquil,
Poeticus, and Polyanthus sections are best, Jonquil
perhaps best of all, for it is without the rather coarse
scent of the Trumpets and Nonsuch, and also escapes
the penetrating lusciousness of *poeticus* and *tazetta*,

which in the south of Europe is exaggerated in the case of *tazetta* into something distinctly unpleasant.

What a delicate refinement there is in the scent of the wild Wood-Violet; it is never overdone. It seems to me to be quite the best of all the violet-scents, just because of its temperate quality. It gives exactly enough, and never that perhaps-just-a-trifle-too-much that may often be noticed about a bunch of frame-Violets, and that also in the south is intensified to a degree that is distinctly undesirable. For just as colour may be strengthened to a painful glare, and sound may be magnified to a torture, so even a sweet scent may pass its appointed bounds and become an overpoweringly evil smell. Even in England several of the Lilies, whose smell is delicious in open-air wafts, cannot be borne in a room. In the south of Europe a Tuberose cannot be brought indoors, and even at home I remember one warm wet August how a plant of Balm of Gilead (*Cedronella triphylla*) had its always powerful but usually agreeably aromatic smell so much exaggerated that it smelt exactly like coal-gas! A brother in Jamaica writes of the large white Jasmine: "It does not do to bring it indoors here; the scent is too strong. One day I thought there was a dead rat under the floor (a thing which did happen once), and behold, it was a glassful of fresh white Jasmine that was the offender!"

While on this less pleasant part of the subject, I cannot help thinking of the horrible smell of the

Dragon Arum; and yet how fitting an accompaniment
it is to the plant, for if ever there was a plant that
looked wicked and repellent, it is this; and yet, like
Medusa, it has its own kind of fearful beauty. In this
family the smell seems to accompany the appearance,
and to diminish in unpleasantness as the flower increases in amiability; for in our native wild Arum the
smell, though not exactly nice, is quite innocuous, and
in the beautiful white Arum or *Calla* of our green-
houses there is as little scent as a flower can well have,
especially one of such large dimensions. In Fungi the
bad smell is nearly always an indication of poisonous
nature, so that it would seem to be given as a warning.
But it has always been a matter of wonder to me why
the root of the harmless and friendly Laurustinus
should have been given a particularly odious smell—a
smell I would rather not attempt to describe. On
moist warmish days in mid-seasons I have sometimes
had a whiff of the same unpleasantness from the bushes
themselves; others of the same tribe have it in a
much lesser degree. There is a curious smell about
the yellow roots of Berberis, not exactly nasty, and a
strong odour, not really offensive, but that I personally
dislike, about the root of *Chrysanthemum maximum*. On
the other hand, I always enjoy digging up, dividing,
and replanting the *Asarums*, both the common Euro-
pean and the American kinds; their roots have a
pleasant and most interesting smell, a good deal like
mild pepper and ginger mixed, but more strongly

aromatic.   The same class of smell, but much fainter,
and always reminding me of very good and delicate
pepper, I enjoy in the flowers of the perennial Lupines.
The only other hardy flowers I can think of whose
smell is distinctly offensive are *Lilium pyrenaicum*,
smelling like a mangy dog, and some of the *Schizanthus*,
that are redolent of dirty hen-house.

There is a class of scent that, though it can neither
be called sweet nor aromatic, is decidedly pleasing and
interesting.   Such is that of Bracken and other Fern-
fronds, Ivy-leaves, Box-bushes, Vine-blossom, Elder-
flowers, and Fig-leaves.   There are the sweet scents
that are wholly delightful—most of the Roses, Honeysuckle, Primrose, Cowslip, Mignonette, Pink, Carnation,
Heliotrope, Lily of the Valley, and a host of others;
then there is a class of scent that is intensely powerful,
and gives an impression almost of intemperance or
voluptuousness, such as Magnolia, Tuberose, Gardenia,
Stephanotis, and Jasmine; it is strange that these all
have white flowers of thick leathery texture.   In
strongest contrast to these are the sweet, wholesome,
wind-wafted scents of clover-field, of bean-field, and of
new-mown hay, and the soft honey-scent of sun-baked
heather, and of a buttercup meadow in April.   Still
more delicious is the wind-swept sweetness of a wood
of Larch or of Scotch Fir, and the delicate perfume of
young-leaved Birch, or the heavier scent of the flowering Lime.   Out on the moorlands, besides the sweet
heather-scent, is that of flowering Broom and Gorse

and of the Bracken, so like the first smell of the sea
as you come near it after a long absence.

How curiously scents of flowers and leaves fall into
classes—often one comes upon related smells running
into one another in not necessarily related plants.
There is a kind of scent that I sometimes meet with
about clumps of Brambles, a little like the waft of a
Fir wood; it occurs again (quite naturally) in the first
taste of blackberry jam, and then turns up again in
Sweet Sultan. It is allied to the smell of the dying
Strawberry leaves.

The smell of the Primrose occurs again in a much
stronger and ranker form in the root-stock, and the
same thing happens with the Violets and Pansies; in
Violets the plant-smell is pleasant, though without
the high perfume of the flower; but the smell of
an overgrown bed of Pansy-plants is rank to offen-
siveness.

Perhaps the most delightful of all flower scents are
those whose tender and delicate quality makes one
wish for just a little more. Such a scent is that of
Apple-blossom, and of some small Pansies, and of the
wild Rose and the Honeysuckle. Among Roses alone
the variety and degree of sweet scent seems almost
infinite. To me the sweetest of all is the Provence,
the old Cabbage Rose of our gardens. When something
approaching this appears, as it frequently does, among
the hybrid perpetuals, I always greet it as the real
sweet Rose smell. One expects every Rose to be

fragrant, and it is a disappointment to find that such a beautiful flower as Baroness Rothschild is wanting in the sweet scent that would be the fitting complement of its incomparable form, and to perceive in so handsome a Rose as Malmaison a heavy smell of decidedly bad quality. But such cases are not frequent.

There is much variety in the scent of the Tea-Roses, the actual tea flavour being strongest in the Dijon class. Some have a powerful scent that is very near that of a ripe Nectarine; of this the best example I know is the old rose Goubault. The half-double red Gloire de Rosamène has a delightful scent of a kind that is rare among Roses. It has a good deal of the quality of that mysterious and delicious smell given off by the dying strawberry leaves, aromatic, pungent, and delicately refined, searching and powerful, and yet subtle and elusive—the best sweet smell of all the year. One cannot have it for the seeking; it comes as it will—a scent that is sad as a forecast of the inevitable certainty of the flower-year's waning, and yet sweet with the promise of its timely new birth.

Sometimes I have met with a scent of somewhat the same mysterious and aromatic kind when passing near a bank clothed with the great St. John's Wort. As this also occurs in early autumn, I suppose it to be occasioned by the decay of some of the leaves. And there is a small yellow-flowered Potentilla that has a scent of the same character, but always freely and willingly given off—a humble-looking little plant, well

worth growing for its sweetness, that much to my
regret I have lost.

I observe that when a Rose exists in both single
and double form the scent is increased in the double
beyond the proportion that one would expect. *Rosa
lucida* in the ordinary single state has only a very
slight scent; in the lovely double form it is very sweet,
and has acquired somewhat of the Moss-rose smell.
The wild Burnet-rose (*R. spinosissima*) has very little
smell; but the Scotch Briars, its garden relatives, have
quite a powerful fragrance, a pale flesh-pink kind,
whose flowers are very round and globe-like, being the
sweetest of all.

But of all the sweet scents of bush or flower, the
ones that give me the greatest pleasure are those of the
aromatic class, where they seem to have a wholesome
resinous or balsamic base, with a delicate perfume
added. When I pick and crush in my hand a twig of
Bay, or brush against a bush of Rosemary, or tread
upon a tuft of Thyme, or pass through incense-laden
brakes of Cistus, I feel that here is all that is best and
purest and most refined, and nearest to poetry, in the
range of faculty of the sense of smell.

The scents of all these sweet shrubs, many of
them at home in dry and rocky places in far-away
lower latitudes, recall in a way far more distinct than
can be done by a mere mental effort of recollection,
rambles of years ago in many a lovely southern land—
in the islands of the Greek Archipelago, beautiful in

form, and from a distance looking bare and arid, and yet with a scattered growth of lowly, sweet-smelling bush and herb, so that as you move among them every plant seems full of sweet sap or aromatic gum, and as you tread the perfumed carpet the whole air is scented; then of dusky groves of tall Cypress and Myrtle, forming mysterious shadowy woodland temples that unceasingly offer up an incense of their own surpassing fragrance, and of cooler hollows in the same lands and in the nearer Orient, where the Oleander grows like the willow of the north, and where the Sweet Bay throws up great tree-like suckers of surprising strength and vigour. It is only when one has seen it grow like this that one can appreciate the full force of the old Bible simile. Then to find oneself standing (while still on earth) in a grove of giant Myrtles fifteen feet high is like having a little chink of the door of heaven opened, as if to show a momentary glimpse of what good things may be beyond!

Among the sweet shrubs from the nearer of these southern regions, one of the best for English gardens is *Cistus laurifolius*. Its wholesome, aromatic sweetness is freely given off, even in winter. In this, as in its near relative, *C. ladaniferus*, the scent seems to come from the gummy surface, and not from the body of the leaf. *Caryopteris Mastacanthus*, the Mastic plant, from China, one of the few shrubs that flower in autumn, has strongly-scented woolly leaves, something like turpentine, but more refined. *Ledum palustre* has a delightful scent when its leaves are bruised. The wild Bog-myrtle, so common in Scotland, has almost the sweet-ness of the true Myrtle, as has also the broad-leaved North American kind, and the Candleberry Gale (*Comptonia asplenifolia*) from the same country. The myrtle-leaved Rhododendron is a dwarf shrub of neat habit, whose bruised leaves have also a myrtle-like smell, though it is less strong than in the Gales. I wonder why the leaves of nearly all the hardy aromatic shrubs are of a hard, dry texture; the exceptions are so few that it seems to be a law.

If my copse were some acres larger I should like nothing better than to make a good-sized clearing, laying out to the sun, and to plant it with these aromatic bushes and herbs. The main planting should be of Cistus and Rosemary and Lavender, and for the shadier edges the Myrtle-leaved Rhododendron, and *Ledum palustre*, and the three Bog-myrtles. Then again in the sun would be Hyssop and Catmint, and Lavender-cotton and Southernwood, with others of the scented Artemisias, and Sage and Marjoram. All the ground would be carpeted with Thyme and Basil and others of the dwarfer sweet-herbs. There would be no regular paths, but it would be so planted that in most parts one would have to brush up against the sweet bushes, and sometimes push through them, as one does on the thinner-clothed of the mountain slopes of southern Italy.

Among the many wonders of the vegetable world

are the flowers that hang their heads and seem to sleep in the daytime, and that awaken as the sun goes down, and live their waking life at night. And those that are most familiar in our gardens have powerful perfumes, except the Evening Primrose (*Œnothera*), which has only a milder sweetness. It is vain to try and smell the night-given scent in the daytime; it is either withheld altogether, or some other smell, quite different, and not always pleasant, is there instead. I have tried hard in daytime to get a whiff of the night sweetness of *Nicotiana affinis*, but can only get hold of something that smells like a horse! Some of the best of the night-scents are those given by the Stocks and Rockets. They are sweet in the hand in the daytime, but the best of the sweet scent seems to be like a thin film on the surface. It does not do to smell them too vigorously, for, especially in Stocks and Wallflowers, there is a strong, rank, cabbage-like under-smell. But in the sweetness given off so freely in the summer evening there is none of this; then they only give their very best.

But of all the family, the finest fragrance comes from the small annual Night-scented Stock (*Matthiola bicornis*), a plant that in daytime is almost ugly; for the leaves are of a dull-grey colour, and the flowers are small and also dull-coloured, and they are closed and droop and look unhappy. But when the sun has set the modest little plant seems to come to life; the grey foliage is almost beautiful in its harmonious relation to the half-light; the flowers stand up and expand, and in the early twilight show tender colouring of faint pink and lilac, and pour out upon the still night-air a lavish gift of sweetest fragrance; and the modest little plant that in strong sunlight looked unworthy of a place in the garden, now rises to its appointed rank and reigns supreme as its prime delight.

CHAPTER XX

THE WORSHIP OF FALSE GODS

SEVERAL times during these notes I have spoken in a disparaging manner of the show-table; and I have not done so lightly, but with all the care and thought and power of observation that my limited capacity is worth; and, broadly, I have come to this: that shows, such as those at the fortnightly meetings of the Royal Horticultural Society, and their more important one in the early summer, whose object is to bring together beautiful flowers of all kinds, to a place where they may be seen, are of the utmost value; and that any shows anywhere for a like purpose, and especially where there are no money prizes, are also sure to be helpful. And the test question I put to myself at any show is this, Does this really help the best interests of horticulture? And as far as I can see that it does this, I think the show right and helpful; and whenever it does not, I think it harmful and misleading.

The love of gardening has so greatly grown and spread within the last few years, that the need of really good and beautiful garden flowers is already far in advance of the demand for the so-called "florists"

flowers, by which I mean those that find favour in
the exclusive shows of Societies for the growing and
exhibition of such flowers as Tulips, Carnations, Dahlias,
and Chrysanthemums.   In support of this I should
like to know what proportion of demand there is, in
Dahlias, for instance, between the show kinds, whose
aim and object is the show-table, and the decorative
kinds, that are indisputably better for garden use.
Looking at the catalogue of a leading Dahlia nursery,
I find that the decorative kinds fill ten pages, while
the show kinds, including Pompones, fill only three.
Is not this some indication of what is wanted in gardens ?

I am of opinion that the show-table is unworthily
used when its object is to be an end in itself, and that
it should be only a means to a better end, and that
when it exhibits what has become merely a "fancy,"
it loses sight of its honourable position as a trustworthy
exponent of horticulture, and has degenerated to a
baser use.   When, as in Chrysanthemum shows, the
flowers on the board are of *no use anywhere but on that
board*, and for the purpose of gaining a money prize, I
hold that the show-table has a debased aim, and a
debasing influence.   Beauty, in all the best sense, is
put aside in favour of set rules and measurements, and
the production of a thing that is of no use or value;
and individuals of a race of plants capable of producing
the highest and most delightful forms of beauty, and
of brightening our homes, and even gardens, during

the dim days of early winter, are teased and tortured
and fatted and bloated into ugly and useless mon-
strosities for no purpose but to gain money. And
when private gardeners go to these shows and see
how the prizes are awarded, and how all the glory is
accorded to the first-prize bloated monster, can we
wonder that the effect on their minds is confusing, if
not absolutely harmful?

Shows of Carnations and Pansies, where the older
rules prevail, are equally misleading, where the single
flowers are arrayed in a flat circle of paper. As with
the Chrysanthemum, every sort of trickery is allowed
in arranging the petals of the Carnation blooms: petals
are pulled out or stuck in, and they are twisted about,
and groomed and combed, and manipulated with
special tools—"dressed," as the show-word has it—
dressed so elaborately that the dressing only stops
short of applying actual paint and perfumery. Already
in the case of Carnations a better influence is being
felt, and at the London shows there are now classes
for border Carnations set up in long-stalked bunches
just as they grow. It is only like this that their
value as outdoor plants can be tested; for many of the
show sorts have miserably weak stalks, and a very
poor, lanky habit of growth.

Then the poor Pansies have single blooms laid flat
on white papers, and are only approved if they will lie
quite flat and show an outline of a perfect circle. All
that is most beautiful in a Pansy, the wing-like curves,

the waved or slightly fluted radiations, the scarcely
perceptible undulation of surface that displays to perfection the admirable delicacy of velvety texture; all
the little tender tricks and ways that make the Pansy
one of the best-loved of garden flowers; all this is
overlooked, and not only passively overlooked, but
overtly contemned. The show-pansy judge appears
to have no eye, or brain, or heart, but to have in
their place a pair of compasses with which to describe
a circle!   All idea of garden delight seems to be
excluded, as this kind of judging appeals to no re-
cognition of beauty for beauty's sake, but to hard
systems of measurement and rigid arrangement and
computation that one would think more applicable to
astronomy or geometry than to any matter relating to
horticulture.

I do most strongly urge that beauty of the highest
class should be the aim, and not anything of the
nature of fashion or "fancy," and that every effort
should be made towards the raising rather than the
lowering of the standard of taste.

The Societies which exist throughout the country
are well organised; many have existed for a great
number of years; they are the local sources of horticultural education, to which large circles of people
naturally look for guidance; and though they produce—
and especially the Rose shows—quantities of beautiful
things, it cannot but be perceived by all who have had
the benefit of some refinement of education, that in

very many cases they either deliberately teach, or at any rate allow to be seen with their sanction, what cannot fail to be debasing to public taste.

I will just take two examples to show how obvious methods of leading taste are not only overlooked, but even perverted; for it is not only in the individual blooms that much of the show-teaching is unworthy, but also in the training of the plants; so that a plant that by nature has some beauty of form, is not encouraged or even allowed to develop that beauty, but is trained into some shape that is not only foreign to its own nature, but is absolutely ugly and ungraceful, and entirely stupid. The natural habit of the Chrysanthemum is to grow in the form of several upright stems. They spring up sheaf-wise, straight upright for a time, and only bending a little outwards above, to give room for the branching heads of bloom. The stems are rather stiff, because they are half woody at the base. In the case of pot-plants it would seem right only so far to stake or train them as to give the necessary support by a few sticks set a little outwards at the top, so that each stem may lean a little over, after the manner of a Bamboo, when their clustered heads of flower would be given enough room, and be seen to the greatest advantage.

But at shows, the triumph of the training art seems to be to drag the poor thing round and round over an internal scaffolding of sticks, with an infinite number of ties and cross-braces, so that it makes a sort of

shapeless ball, and to arrange the flowers so that they
are equally spotted all over it, by tying back some almost
to snapping-point, and by dragging forward others to the
verge of dislocation.   I have never seen anything so
ugly in the way of potted plants as a certain kind of
Chrysanthemum that has incurved flowers of heavy
sort of dull leaden-looking red-purple colour trained
in this manner.   Such a sight gives me a feeling of
shame, not unmixed with wrathful indignation.   I ask
myself, What is it for ? and I get no answer.   I ask
a practical gardener what it is for, and he says, " Oh,
it is one of the ways they are trained for shows."   I
ask him, Does he think it pretty, or is it any use ? and
he says, " Well, they think it makes a nice variety ; "
and when I press him further, and say I consider it a
very nasty variety, and does he think nasty varieties are
better than none, the question is beyond him, and he
smiles vaguely and edges away, evidently thinking my
conversation perplexing, and my company undesirable.
I look again at the unhappy plant, and see its poor
leaves fat with an unwholesome obesity, and seeming
to say, We were really a good bit mildewed, but have
been doctored up for the show by being crammed and
stuffed with artificial aliment !

My second example is that of *Azalea indica*.   What
is prettier in a room than one of these in its little tree
form, a true tree, with tiny trunk and wide-spreading
branches, and its absurdly large and lovely flowers ?
Surely it is the most perfect room ornament that we

can have in tree shape in a moderate-sized pot; and
where else can one see a tree loaded with lovely bloom
whose individual flowers have a diameter equal to five
times that of the trunk?

But the show decrees that all this is wrong, and
that the tiny, brittle branches must be trained stiffly
round till the shape of the plant shows as a sort of
cylinder. Again I ask myself, What is this for? What
does it teach? Can it be really to teach with
deliberate intention that instead of displaying its
natural and graceful tree form it should aim at a more
desirable kind of beauty, such as that of the chimney-
pot or drain-pipe, and that this is so important that it
is right and laudable to devote to it much time and
delicate workmanship?

I cannot but think, as well as hope, that the strong
influences for good that are now being brought to bear
on all departments of gardening may reach this class of
show, for there are already more hopeful signs in the
admission of classes for groups arranged for decoration.

The prize-show system no doubt creates its own
evils, because the judges, and those who frame the
schedules, have been in most cases men who have a
knowledge of flowers, but who are not people of cultivated taste, and in deciding what points are to consti-
tute the merits of a flower they have to take such
qualities as are within the clearest understanding of
people of average intelligence and average education—
such, for instance, as size that can be measured,

symmetry that can be easily estimated, thickness of petal that can be felt, and such qualities of colour as appeal most strongly to the uneducated eye; so that a flower may possess features or qualities that endow it with the highest beauty, but that exclude it, because the hard and narrow limits of the show-laws provide no means of dealing with it. It is, therefore, thrown out, not because they have any fault to find with it, but because it does not concern them; and the ordinary gardener, to whose practice it might be of the highest value, accepting the verdict of the show-judge as an infallible guide, also treats it with contempt and neglect.

Now, all this would not so much matter if it did not delude those whose taste is not sufficiently educated to enable them to form an opinion of their own in accordance with the best and truest standards of beauty; for I venture to repeat that what we have to look for for the benefit of our gardens, and for our own bettering and increase of happiness in those gardens, are things that are beautiful, rather than things that are round, or straight, or thick, still less than for those that are new, or curious, or astonishing. For all these false gods are among us, and many are they who are willing to worship.

CHAPTER XXI

NOVELTY AND VARIETY

WHEN I look back over thirty years of gardening, I see what an extraordinary progress there has been, not only in the introduction of good plants new to general cultivation, but also in the home production of improved kinds of old favourites. In annual plants alone there has been a remarkable advance. And here again, though many really beautiful things are being brought forward, there seems always to be an undue value assigned to a fresh development, on the score of its novelty.

Now it seems to me, that among the thousands of beautiful things already at hand for garden use, there is no merit whatever in novelty or variety unless the thing new or different is distinctly more beautiful, or in some such way better than an older thing of the same class.

And there seems to be a general wish among seed growers just now to dwarf all annual plants. Now, when a plant is naturally of a diffuse habit, the fixing of a dwarfer variety may be a distinct gain to horticulture—it may just make a good garden plant out of one that was formerly of indifferent quality; but there seems to me to be a kind of stupidity in inferring from this that all annuals are the better for dwarfing. I take it that the bedding system has had a good deal to do with it.   It no doubt enables ignorant gardeners to use a larger variety of plants as senseless colour-masses, but it is obvious that many, if not most, of the plants are individually made much uglier by the process.   Take, for example, one of the dwarfest Agera-tums: what a silly little dumpy, formless, pincushion of a thing it is!   And then the dwarfest of the China Asters.   Here is a plant (whose chief weakness already lies in a certain over-stiffness) made stiffer and more shapeless still by dwarfing and by cramming with too many petals.   The Comet Asters of later years are a much-improved type of flower, with a looser shape and a certain degree of approach to grace and beauty. When this kind came out it was a noteworthy novelty, not because it was a novelty, but because it was a better and more beautiful thing.   Also among the same Asters the introduction of a better class of red colouring, first of the blood-red and then of the so-called scarlet shades, was a good variety, because it was the distinct bettering of the colour of a popular race of garden-flowers, whose red and pink colourings had hitherto been of a bad and rank quality.

It is quite true that here and there the dwarf kind is a distinctly useful thing, as in the dwarf Nasturtiums.   In this grand plant one is glad to have

MULLEINS GROWING IN THE FACE OF DRY WALL.

dwarf ones as well as the old trailing kinds. I even confess to a certain liking for the podgy little dwarf Snapdragons; they are ungraceful little dumpy things, but they happen to have come in some tender colourings of pale yellow and pale pink, that give them a kind of absurd prettiness, and a certain garden-value. I also look at them as a little floral joke that is harmless and not displeasing, but they cannot for a moment compare in beauty with the free-growing Snapdragon of the older type. This I always think one of the best and most interesting and admirable of garden-plants. Its beauty is lost if it is crowded up among other things in a border; it should be grown in a dry wall or steep rocky bank, where its handsome bushy growth and finely-poised spikes of bloom can be well seen.

One of the annuals that I think is entirely spoilt by dwarfing is Love-in-a-Mist, a plant I hold in high admiration. Many years ago I came upon some of it in a small garden, of a type that I thought extremely desirable, with a double flower of just the right degree of fulness, and of an unusually fine colour. I was fortunate enough to get some seed, and have never grown any other, nor have I ever seen elsewhere any that I think can compare with it.

The Zinnia is another fine annual that has been much spoilt by its would-be improvers. When a Zinnia has a hard, stiff, tall flower, with a great many rows of petals piled up one on top of another, and

when its habit is dwarfed to a mean degree of squat-
ness, it looks to me both ugly and absurd, whereas
a reasonably double one, well branched, and two feet
high, is a handsome plant.

I also think that Stocks and Wallflowers are much
handsomer when rather tall and branching. Dwarf
Stocks, moreover, are invariably spattered with soil in
heavy autumn rain.

An example of the improver not knowing where to
stop in the matter of colouring, always strikes me in
the Gaillardias, and more especially in the perennial
kind, that is increased by division as well as by seed.
The flower is naturally of a strong orange-yellow colour,
with a narrow ring of red round the centre. The
improver has sought to increase the width of the red
ring. Up to a certain point it makes a livelier and
brighter-looking flower; but he has gone too far, and
extended the red till it has become a red flower with
a narrow yellow edge. The red also is of a rather
dull and heavy nature, so that instead of a handsome
yellow flower with a broad central ring, here is an ugly
red one with a yellow border. There is no positive
harm done, as the plant has been propagated at every
stage of development, and one may choose what one
will; but to see them together is an instructive lesson.

No annual plant has of late years been so much
improved as the Sweet Pea, and one reason why its
charming beauty and scent are so enjoyable is, that
they grow tall, and can be seen on a level with the

eye. There can be no excuse whatever for dwarfing
this, as has lately been done. There are already
plenty of good flowering plants under a foot high, and
the little dwarf white monstrosity, now being followed
by coloured ones of the same habit, seems to me
worthy of nothing but condemnation.   It would be
as right and sensible to dwarf a Hollyhock into a
podgy mass a foot high, or a Pentstemon, or a Foxglove. Happily these have as yet escaped dwarfing,
though I regret to see that a deformity that not un-
frequently appears among garden Foxgloves, looking
like a bell-shaped flower topping a stunted spike,
appears to have been " fixed," and is being offered as
a " novelty." Here is one of the clearest examples of
a new development which is a distinct debasement of
a naturally beautiful form, but which is nevertheless
being pushed forward in trade : it has no merit whatever in itself, and is only likely to sell because it is
new and curious.

And all this parade of distortion and deformity
comes about from the grower losing sight of beauty as
the first consideration, or from his not having the
knowledge that would enable him to determine what
are the points of character in various plants most
deserving of development, and in not knowing when
or where to stop.   Abnormal size, whether greatly
above or much below the average, appeals to the vulgar
and uneducated eye, and will always command its
attention and wonderment.   But then the production

of the immense size that provokes astonishment, and
the misapplied ingenuity that produces unusual dwarfing, are neither of them very high aims.

And much as I feel grateful to those who improve
garden flowers, I venture to repeat my strong conviction that their efforts in selection and other methods
should be so directed as to keep in view the attainment
of beauty in the first place, and as a point of honour;
not in mere increase of size of bloom or compactness
of habit—many plants have been spoilt by excess of
both; not for variety or novelty as ends in themselves,
but only to welcome them, and offer them, if they are
distinctly of garden value in the best sense.   For if
plants are grown or advertised or otherwise pushed on
any other account than that of their possessing some
worthy form of beauty, they become of the same nature
as any other article in trade that is got up for sale
for the sole benefit of the seller, that is unduly lauded
by advertisement, and that makes its first appeal to
the vulgar eye by an exaggerated and showy pictorial
representation; that will serve no useful purpose, and
for which there is no true or healthy demand.

No doubt much of it comes about from the unwholesome pressure of trade competition, which in a
way obliges all to follow where some lead. I trust
that my many good friends in the trade will understand that my remarks are not made in any personal
sense whatever.   I know that some of them feel much
as I do on some of these points, but that in many

ways they are helpless, being all bound in a kind of bondage to the general system. And there is one great evil that calls loudly for redress, but that will endure until some of the mightiest of them have the energy and courage to band themselves together and to declare that it shall no longer exist among them.

CHAPTER XXII

WEEDS AND PESTS

WEEDING is a delightful occupation, especially after summer rain, when the roots come up clear and clean. One gets to know how many and various are the ways of weeds—as many almost as the moods of human creatures. How easy and pleasant to pull up are the soft annuals like Chickweed and Groundsel, and how one looks with respect at deep-rooted things like Docks, that make one go and fetch a spade. Comfrey is another thing with a terrible root, and every bit must be got out, as it will grow again from the smallest scrap. And hard to get up are the two Bryonies, the green and the black, with such deep-reaching roots, that, if not weeded up within their first year, will have to be seriously dug out later. The white Convolvulus, one of the loveliest of native plants, has a most persistently running root, of which every joint will quickly form a new plant. Some of the worst weeds to get out are Goutweed and Coltsfoot. Though I live on a light soil, comparatively easy to clean, I have done some gardening in clay, and well know what

a despairing job it is to get the bits of either of these
roots out of the stiff clods.

The most persistent weed in my soil is the small
running Sheep's Sorrel. First it makes a patch, and
then sends out thready running roots all round, a foot
or more long; these, if not checked, establish new
bases of operation, and so it goes on, always spreading
farther and farther. When this happens in soft ground
that can be hoed and weeded it matters less, but in
the lawn it is a more serious matter. Its presence
always denotes a poor, sandy soil of rather a sour
quality.

Goutweed is a pest in nearly all gardens, and very
difficult to get out. When it runs into the root of
some patch of hardy plant, if the plant can be spared,
I find it best to send it at once to the burn-heap; or
if it is too precious, there is nothing for it but to cut it
all up and wash it out, to be sure that not the smallest
particle of the enemy remains. Some weeds are
deceiving—Sow-thistle, for instance, which has the look
of promising firm hand-hold and easy extraction, but
has a disappointing way of almost always breaking
short off at the collar. But of all the garden weeds
that are native plants I know none so persistent or
so insidious as the Rampion Bell-flower (*Campanula
Rapunculus*); it grows from the smallest thread of root,
and it is almost impossible to see every little bit; for
though the main roots are thick, and white, and fleshy,
the fine side roots that run far abroad are very small,

R

and of a reddish colour, and easily hidden in the brown earth.

But some of the worst garden-weeds are exotics run wild. The common Grape Hyacinth sometimes overruns a garden and cannot be got rid of. *Sambucus ebulis* is a plant to beware of, its long thong-like roots spreading far and wide, and coming up again far away from the parent stock. For this reason it is valuable for planting in such places as newly-made pond-heads, helping to tie the bank together. *Polygonum Sieboldi* must also be planted with caution. The winter Heliotrope (*Petasites fragrans*) is almost impossible to get out when once it has taken hold, growing in the same way as its near relative the native Coltsfoot.

But by far the most difficult plant to abolish or even keep in check that I know is *Ornithogalum nutans*. Beautiful as it is, and valuable as a cut flower, I will not have it in the garden. I think I may venture to say that in this soil, when once established, it cannot be eradicated. Each mature bulb makes a host of offsets, and the seed quickly ripens. When it is once in a garden it will suddenly appear in all sorts of different places. It is no use trying to dig it out. I have dug out the whole space of soil containing the patch, a barrow-load at a time, and sent it to the middle of the burn-heap, and put in fresh soil, and there it is again next year, nearly as thick as ever. I have dug up individual small patches with the greatest care, and got out every bulb and offset, and every bit of the whitish leaf stem, for I have such faith in its power of reproduction that I think every atom of this is capable of making a plant, only to find next year a thriving young tuft of the "grass" in the same place. And yet the bulb and underground stem are white, and the earth is brown, and I passed it all several times through my fingers, but all in vain. I confess that it beats me entirely.

Coronilla varia is a little plant that appears in catalogues among desirable Alpines, but is a very "rooty" and troublesome thing, and scarcely good enough for garden use, though pretty in a grassy bank where its rambling ways would not be objectionable. I once brought home from Brittany some roots of *Linaria repens*, that looked charming by a roadside, and planted them in a bit of Alpine garden, a planting that I never afterwards ceased to regret.

I learnt from an old farmer a good way of getting rid of a bed of nettles—to thrash them down with a stick every time they grow up. If this is done about three times during the year, the root becomes so much weakened that it is easily forked out, or if the treatment is gone on with, the second year the nettles die. Thrashing with a stick is better than cutting, as it makes the plant bleed more; any mutilation of bruise or ragged tearing of fibre is more harmful to plant or tree than clean cutting.

Of bird, beast, and insect pests we have plenty. First, and worst, are rabbits. They will gnaw and nibble anything and everything that is newly planted, even native things like Juniper, Scotch Fir, and Gorse. The necessity of wiring everything newly planted adds greatly to the labour and expense of the garden, and the unsightly grey wire-netting is an unpleasant eyesore. When plants or bushes are well established the rabbits leave them alone, though some families of plants are always irresistible—Pinks and Carnations, for instance, and nearly all Cruciferæ, such as Wallflowers, Stocks, and Iberis. The only plants I know that they do not touch are Rhododendrons and Azaleas; they leave them for the hare, that is sure to get in every now and then, and who stands up on his long hind-legs, and will eat Rose-bushes quite high up.

Plants eaten by a hare look as if they had been cut with a sharp knife; there is no appearance of gnawing or nibbling, no ragged edges of wood or frayed bark, but just a straight clean cut.

Field mice are very troublesome. Some years they will nibble off the flower-buds of the Lent Hellebores; when they do this they have a curious way of collecting them and laying them in heaps. I have no idea why they do this, as they neither carry them away nor eat them afterwards; there the heaps of buds lie till they rot or dry up. They once stole all my Auricula seed in the same way. I had marked some good plants for seed, cutting off all the other flowers as soon as they went out of bloom. The seed was ripening, and I watched it daily, awaiting the

moment for harvesting.   But a few days before it was
ready I went round and found the seed was all gone;
it had been cut off at the top of the stalk, so that the
umbel-shaped heads had been taken away whole.   I
looked about, and luckily found three slightly hollow
places under the bank at the back of the border where
the seed-heads had been piled in heaps.   In this case it
looked as if it had been stored for food; luckily it was
near enough to ripeness for me to save my crop.

The mice are also troublesome with newly-sown
Peas, eating some underground, while sparrows nibble
off others when just sprouted; and when outdoor Grapes
are ripening mice run up the walls and eat them.
Even when the Grapes are tied in oiled canvas bags
they will eat through the bags to get at them, though I
have never known them to gnaw through the news-
paper bags that I now use in preference, and that
ripen the Grapes as well.   I am not sure whether it is
mice or birds that pick off the flowers of the big bunch
Primroses, but am inclined to think it is mice, because
the stalks are cut low down.

Pheasants are very bad gardeners; what they seem
to enjoy most are Crocuses—in fact, it is no use planting
them.   I had once a nice collection of Crocus species.
They were in separate patches, all along the edge of one
border, in a sheltered part of the garden, where pheasants did not often come.   One day when I came to
see my Crocuses, I found where each patch had been a
basin-shaped excavation and a few fragments of stalk

or some part of the plant. They had begun at one end and worked steadily along, clearing them right out. They also destroyed a long bed of *Anemone fulgens*. First they took the flowers, and then the leaves, and lastly pecked up and ate the roots.

But we have one grand consolation in having no slugs, at least hardly any that are truly indigenous; they do not like our dry, sandy heaths. Friends are very generous in sending them with plants, so that we have a moderate number that hang about frames and pot plants, though nothing much to boast of; but they never trouble seedlings in the open ground, and for this I can never be too thankful.

Alas that the beautiful bullfinch should be so dire an enemy to fruit-trees, and also the pretty little tits! but so it is; and it is a sad sight to see a well-grown fruit-tree with all its fruit-buds pecked out and lying under it on the ground in a thin green carpet. We had some fine young cherry-trees in a small orchard that we cut down in despair after they had been growing twelve years. They were too large to net, and their space could not be spared just for the mischievous fun of the birds.

CHAPTER XXIII

THE BEDDING FASHION AND ITS INFLUENCE

IT is curious to look back at the old days of bedding-out, when that and that only meant gardening to most people, and to remember how the fashion, beginning in the larger gardens, made its way like a great inundating wave, submerging the lesser ones, and almost drowning out the beauties of the many little flowery cottage plots of our English waysides. And one wonders how it all came about, and why the bedding system, admirable for its own purpose, should have thus outstepped its bounds, and have been allowed to run riot among gardens great and small throughout the land. But so it was, and for many years the fashion, for it was scarcely anything better, reigned supreme.

It was well for all real lovers of flowers when some quarter of a century ago a strong champion of the good old flowers arose, and fought strenuously to stay the devastating tide, and to restore the healthy liking for the good old garden flowers. Many soon followed, and now one may say that all England has flocked to the standard. Bedding as an all-prevailing fashion is now dead; the old garden-flowers are again honoured

and loved, and every encouragement is freely offered
to those who will improve old kinds and bring forward
others.

And now that bedding as a fashion no longer exists,
one can look at it more quietly and fairly, and see
what its uses really are, for in its own place and way
it is undoubtedly useful and desirable. Many great
country-houses are only inhabited in winter, then perhaps for a week or two at Easter, and in the late
summer. There is probably a house-party at Easter,
and a succession of visitors in the late summer. A
brilliant garden, visible from the house, dressed for
spring and dressed for early autumn, is exactly what
is wanted—not necessarily from any special love of
flowers, but as a kind of bright and well-kept furnishing of the immediate environment of the house. The
gardener delights in it; it is all routine work; so
many hundreds or thousands of scarlet Geranium, of
yellow Calceolaria, of blue Lobelia, of golden Feverfew,
or of other coloured material. It wants no imagina-
tion; the comprehension of it is within the range of
the most limited understanding; indeed its prevalence
for some twenty years or more must have had a
deteriorating influence on the whole class of private
gardeners, presenting to them an ideal so easy of
attainment and so cheap of mental effort.

But bedding, though it is gardening of the least
poetical or imaginative kind, can be done badly or
beautifully. In the *parterre* of the formal garden it

is absolutely in place, and brilliantly-beautiful pictures can be made by a wise choice of colouring. I once saw, and can never forget, a bedded garden that was a perfectly satisfying example of colour-harmony; but then it was planned by the master, a man of the most refined taste, and not by the gardener. It was a *parterre* that formed part of the garden in one of the fine old places in the Midland counties. I have no distinct recollection of the design, except that there was some principle of fan-shaped radiation, of which each extreme angle formed one centre. The whole garden was treated in one harmonious colouring of full yellow, orange, and orange-brown, half-hardy annuals, such as French and African Marigolds, Zinnias, and Nasturtiums, being freely used. It was the most noble treatment of one limited range of colouring I have ever seen in a garden; brilliant without being garish, and sumptuously gorgeous without the reproach of gaudiness—a precious lesson in temperance and restraint in the use of the one colour, and an admirable exposition of its powerful effect in the hands of a true artist.

I think that in many smaller gardens a certain amount of bedding may be actually desirable; for where the owner of a garden has a special liking for certain classes or mixtures of plants, or wishes to grow them thoroughly well and enjoy them individually to the full, he will naturally grow them in separate beds, or may intentionally combine the beds, if he will, into some form of good garden effect. But the great fault of the bedding system when at its height was, that it swept over the country as a tyrannical fashion, that demanded, and for the time being succeeded in effecting, the exclusion of better and more thoughtful kinds of gardening; for I believe I am right in saying that it spread like an epidemic disease, and raged far and wide for nearly a quarter of a century.

Its worst form of all was the "ribbon border," generally a line of scarlet Geranium at the back, then a line of Calceolaria, then a line of blue Lobelia, and lastly, a line of the inevitable Golden Feather Feverfew, or what our gardener used to call Featherfew. Could anything be more tedious or more stupid? And the ribbon border was at its worst when its lines were not straight, but waved about in weak and silly sinuations.

And when bedding as a fashion was dead, when this false god had been toppled off his pedestal, and his worshippers had been converted to better beliefs, in turning and rending him they often went too far, and did injustice to the innocent by professing a dislike to many a good plant, and renouncing its use. It was not the fault of the Geranium or of the Calceolaria that they had been grievously misused and made to usurp too large a share of our garden spaces. Not once but many a time my visitors have expressed unbounded surprise when they saw these plants in my garden, saying, "I should have thought that you

GERANIUMS IN NEAPOLITAN POTS.

would have despised Geraniums." On the contrary, I love Geraniums. There are no plants to come near them for pot, or box, or stone basket, or for massing in any sheltered place in hottest sunshine; and I love their strangely - pleasant smell, and their beautiful modern colourings of soft scarlet and salmon-scarlet and salmon-pink, some of these grouping beautifully together. I have a space in connection with some formal stonework of steps, and tank, and paved walks, close to the house, on purpose for the summer placing of large pots of Geranium, with sometimes a few Cannas and Lilies. For a quarter of the year it is one of the best things in the garden, and delightful in colour. Then no plant does so well or looks so suitable in some earthen pots and boxes from Southern Italy that I always think the best that were ever made, their shape and well-designed ornament traditional from the Middle Ages, and probably from an even more remote antiquity.

There are, of course, among bedding Geraniums many of a bad, raw quality of colour, particularly among cold, hard pinks, but there are so many to choose from that these can easily be avoided.

I remember some years ago, when the bedding fashion was going out, reading some rather heated discussions in the gardening papers about methods of planting out and arranging various tender but indispensable plants. Some one who had been writing about the errors of the bedding system wrote about

planting some of these in isolated masses. He was
pounced upon by another, who asked, "What is this
but bedding?" The second writer was so far justified,
in that it cannot be denied that any planting in beds
is bedding. But then there is bedding and bedding—
a right and a wrong way of applying the treatment.
Another matter that roused the combative spirit of
the captious critic was the filling up of bare spaces
in mixed borders with Geraniums, Calceolarias, and
other such plant. Again he said, "What is this but
bedding? These are bedding plants." When I read
this it seemed to me that his argument was, These
plants may be very good plants in themselves, but
because they have for some years been used wrongly,
therefore they must not now be used rightly! In the
case of my own visitors, when they have expressed
surprise at my having "those horrid old bedding
plants" in my gárden, it seemed quite a new view
when I pointed out that bedding plants were only pas-
sive agents in their own misuse, and that a Geranium
was a Geranium long before it was a bedding plant!
But the discussion raised in my mind a wish to come
to some conclusion about the difference between bedding in the better and worse sense, in relation to the
cases quoted, and it appeared to me to be merely in
the choice between right and wrong placing—placing
monotonously or stupidly, so as merely to fill the space,
or placing with a feeling for "drawing" or proportion.
For I had very soon found out that, if I had a number

SPACE IN STEP AND TANK-GARDEN FOR LILIES, CANNAS, AND GERANIUMS.

HYDRANGEAS IN TUBS, IN A PART OF THE SAME GARDEN.

of things to plant anywhere, whether only to fill up
a border or as a detached group, if I placed the
things myself, carefully exercising what power of dis-
crimination I might have acquired, it looked fairly
right, but that if I left it to one of my garden people
(a thing I rarely do) it looked all nohow, or like bedding in the worst sense of the word.

Even the better ways of gardening do not wholly
escape the debasing influence of fashion. Wild gardening is a delightful, and in good hands a most desirable,
pursuit, but no kind of gardening is so difficult to do
well, or is so full of pitfalls and of paths of peril.
Because it has in some measure become fashionable,
and because it is understood to mean the planting of
exotics in wild places, unthinking people rush to the
conclusion that they can put any garden plants into any
wild places, and that that is wild gardening. I have seen
woody places that were already perfect with their own
simple charm just muddled and spoilt by a reckless
planting of garden refuse, and heathy hillsides already
sufficiently and beautifully clothed with native vegetation made to look lamentably silly by the planting of
a nurseryman's mixed lot of exotic Conifers.

In my own case, I have always devoted the
most careful consideration to any bit of wild gardening I thought of doing, never allowing myself
to decide upon it till I felt thoroughly assured that
the place seemed to ask for the planting in contem-
plation, and that it would be distinctly a gain in

pictorial value; so there are stretches of Daffodils in one part of the copse, while another is carpeted with Lily of the Valley. A cool bank is covered with Gaultheria, and just where I thought they would look well as little jewels of beauty, are spreading patches of Trillium and the great yellow Dog-tooth Violet. Besides these there are only some groups of the Giant Lily. Many other exotic plants could have been made to grow in the wooded ground, but they did not seem to be wanted; I thought where the copse looked well and complete in itself it was better left alone.

But where the wood joins the garden some bold groups of flowering plants are allowed, as of Mullein in one part and Foxglove in another; for when standing in the free part of the garden, it is pleasant to project the sight far into the wood, and to let the garden influences penetrate here and there, the better to join the one to the other.

MULLEIN (VERBASCUM PHLOMOIDES) AT THE EDGE
OF THE FIR WOOD.

A GRASS PATH IN THE COPSE. (*See page* 61.)

Under the Bracken in both pictures is a wide planting of Lily of the Valley, flowering in May before the Fern is up.

CHAPTER XXIV

MASTERS AND MEN

Now that the owners of good places are for the most part taking a newly-awakened and newly-educated pleasure in the better ways of gardening, a frequent source of difficulty arises from the ignorance and obstructiveness of gardeners. The owners have become aware that their gardens may be sources of the keenest pleasure. The gardener may be an excellent man, perfectly understanding the ordinary routine of garden work; he may have been many years in his place; it is his settled home, and he is getting well on into middle life; but he has no understanding of the new order of things, and when the master, perfectly understanding what he is about, desires that certain things shall be done, and wishes to enjoy the pleasure of directing the work himself, and seeing it grow under his hand, he resents it as an interference, and becomes obstructive, or does what is required in a spirit of such sullen acquiescence that it is equal to open opposition. And I have seen so many gardens and gardeners that I have come to recognise certain types; and this one, among men of a certain age, is unfortunately frequent.

Various degrees of ignorance and narrow-mindedness must no doubt be expected among the class that produces private gardeners. Their general education is not very wide to begin with, and their training is usually all in one groove, and the many who possess a full share of vanity get to think that, because they have exhausted the obvious sources of experience that have occurred within their reach, there is nothing more to learn, or to know, or to see, or to feel, or to enjoy. It is in this that the difficulty lies. The man has no doubt done his best through life; he has performed his duties well and faithfully, and can render a good account of his stewardship. It is no fault of his that more means of enlarging his mind have not been within his grasp, and, to a certain degree, he may be excused for not understanding that there is anything beyond; but if he is naturally vain and stubborn his case is hopeless. If, on the other hand, he is wise enough to know that he does not know everything, and modest enough to acknowledge it, as do all the greatest and most learned of men, he will then be eager to receive new and enlarged impressions, and his willing and intelligent co-operation will be a new source of interest in life both to himself and his employer, as well as a fresh spring of vitality in the life of the garden. I am speaking of the large middle class of private gardeners, not of those of the highest rank, who have among them men of good education and a large measure of refinement. From among these I

think of the late Mr. Ingram of the Belvoir Castle
gardens with regret, as for a personal friend, and also
as of one who was a true garden artist.

But most people who have fair-sized gardens have
to do with the middle class of gardener, the man of
narrow mental training. The master who, after a good
many years of active life, is looking forward to settling
in his home and improving and enjoying his garden,
has had so different a training, a course of teaching so
immeasurably wider and more enlightening. As a boy
he was in a great public school, where, by wholesome
friction with his fellows, he had any petty or personal
nonsense knocked out of him while still in his early
"teens." Then he goes to college, and whether stu-
diously inclined or not, he is already in the great
world, always widening his ideas and experience. Then
perhaps he is in one of the active professions, or
engaged in scientific or intellectual research, or in
diplomacy, his ever-expanding intelligence rubbing up
against all that is most enlightened and astute in men,
or most profoundly inexplicable in matter. He may
be at the same time cultivating his taste for literature
and the fine arts, searching the libraries and galleries
of the civilised world for the noblest and most divinely-
inspired examples of human work, seeing with an
eye that daily grows more keenly searching, and re-
ceiving and holding with a brain that ever gains a
firmer grasp, and so acquires some measure of the
higher critical faculty. He sees the ruined gardens of

antiquity, colossal works of the rulers of Imperial Rome, and the later gardens of the Middle Ages (direct descendants of those greater and older ones), some of them still among the most beautiful gardens on earth. He sees how the taste for gardening grew and travelled, spreading through Europe and reaching England, first, no doubt, through her Roman invaders. He becomes more and more aware of what great and enduring happiness may be enjoyed in a garden, and how all that he can learn of it in the leisure intervals of his earlier maturity, and then in middle life, will help to brighten his later days, when he hopes to refine and make better the garden of the old home by a reverent application of what he has learnt. He thinks of the desecrated old bowling-green, cut up to suit the fashion of thirty years ago into a patchwork of incoherent star and crescent shaped beds; of how he will give it back its ancient character of unbroken repose; he thinks how he will restore the string of fish-ponds in the bottom of the wooded valley just below, now a rushy meadow with swampy hollows that once were ponds, and humpy mounds, ruins of the ancient dikes; of how the trees will stand reflected in the still water; and how he will live to see again in middle hours of summer days, as did the monks of old, the broad backs of the golden carp basking just below the surface of the sun-warmed water.

And such a man as this comes home some day and finds the narrow-minded gardener, who believes that he already knows all that can be known about gardening, who thinks that the merely technical part, which he perfectly understands, is all that there is to be known and practised, and that his crude ideas about arrangement of flowers are as good as those of any one else. And a man of this temperament cannot be induced to believe, and still less can he be made to understand, that all that he knows is only the means to a further and higher end, and that what he can show of a completed garden can only reach to an average dead-level of dulness compared with what may come of the life-giving influence of one who has the mastery of the higher garden knowledge.

Moreover, he either forgets, or does not know, what is the main purpose of a garden, namely, that it is to give its owner the best and highest kind of earthly pleasure. Neither is he enlightened enough to understand that the master can take a real and intelligent interest in planning and arranging, and in watching the working out in detail. His small-minded vanity can only see in all this a distrust in his own powers and an intentional slight in his own ability, whereas no such idea had ever entered the master's mind.

Though there are many of this kind of gardener (and with their employers, if they have the patience to retain them in their service, I sincerely condole), there are happily many of a widely-different nature, whose minds are both supple and elastic and intelligently receptive, who are eager to learn and to try what has not yet come within the range of their experience, who show a cheerful readiness to receive a fresh range of ideas, and a willing alacrity in doing their best to work them out. Such a servant as this warms his master's heart, and it would do him good to hear, as I have many times heard, the terms in which the master speaks of him. For just as the educated man feels contempt for the vulgar pretension that goes with any exhibition of ignorant vanity, so the evidence of the higher qualities commands his respect and warm appreciation. Among the gardeners I have known, five such men come vividly to my recollection—good men all, with a true love of flowers, and its reflection of happiness written on their kindly faces.

But then, on the other hand, frequent causes of irritation arise between master and man from the master's ignorance and unreasonable demands. For much as the love of gardening has grown of late, there are many owners who have no knowledge of it whatever. I have more than once had visitors who complained of their gardeners, as I thought quite unreasonably, on their own showing. For it is not enough to secure the services of a thoroughly able man, and to pay good wages, and to provide every sort of appliance, if there is no reasonable knowledge of what it is right and just to expect. I have known a lady, after paying a round of visits in great houses, complain of her gardener. She had seen at one place remarkably fine forced strawberries, at another some phenomenal frame Violets, and at a third immense

Malmaison Carnations; whereas her own gardener did not excel in any of these, though she admitted that he was admirable for Grapes and Chrysanthemums. "If the others could do all these things to perfection," she argued, "why could not he do them?" She expected her gardener to do equally well all that she had seen best done in the other big places. It was in vain that I pleaded in defence of her man that all gardeners were human creatures, and that it was in the nature of such creatures to have individual aptitudes and special preferences, and that it was to be expected that each man should excel in one thing, or one thing at a time, and so on; but it was of no use, and she would not accept any excuse or explanation.

I remember another example of a visitor who had a rather large place, and a gardener who had as good a knowledge of hardy plants as one could expect. My visitor had lately got the idea that he liked hardy flowers, though he had scarcely thrown off the influence of some earlier heresy which taught that they were more or less contemptible—the sort of thing for cottage gardens; still, as they were now in fashion, he thought he had better have them. We were passing along my flower-border, just then in one of its best moods of summer beauty, and when its main occupants, three years planted, had come to their full strength, when, speaking of a large flower-border he had lately had made, he said, "I told my fellow last autumn to get anything he liked, and yet it is perfectly wretched.

It is not as if I wanted anything out of the way; I only want a lot of common things like that," waving a hand airily at my precious border, while scarcely taking the trouble to look at it.

And I have had another visitor of about the same degree of appreciative insight, who, contemplating some cherished garden picture, the consummation of some long-hoped-for wish, the crowning joy of years of labour, said, " Now look at that; it is just right, and yet it is quite simple—there is absolutely nothing in it; now, why can't my man give me that ? "

I am far from wishing to disparage or undervalue the services of the honest gardener, but I think that on this point there ought to be the clearest understanding; that the master must not expect from the gardener accomplishments that he has no means of acquiring, and that the gardener must not assume that his knowledge covers all that can come within the scope of the widest and best practice of his craft. There are branches of education entirely out of his reach that can be brought to bear upon garden planning and arrangement down to the very least detail. What the educated employer who has studied the higher forms of gardening can do or criticise, he cannot be expected to do or understand; it is in itself almost the work of a lifetime, and only attainable, like success in any other fine art, by persons of, firstly, special temperament and aptitude; and, secondly, by their unwearied study and closest application.

But the result of knowledge so gained shows itself throughout the garden. It may be in so simple a thing as the placing of a group of plants. They can be so placed by the hand that knows, that the group is in perfect drawing in relation to what is near; while by the ordinary gardener they would be so planted that they look absurd, or unmeaning, or in some way awkward and unsightly. It is not enough to cultivate plants well; they must also be used well. The servant may set up the canvas and grind the colours, and even set the palette, but the master alone can paint the picture. It is just the careful and thoughtful exercise of the higher qualities that makes a garden interesting, and their absence that leaves it blank, and dull, and lifeless. I am heartily in sympathy with the feeling described in these words in a friend's letter, " I think there are few things so interesting as to see in what way a person, whose perceptions you think fine and worthy of study, will give them expression in a garden."

INDEX

THE END

Printed by BALLANTYNE, HANSON & Co.
Edinburgh & London

Printed in the United States
By Bookmasters